Quercus

VOYAGE
INTO THE
DEEP

SALLY MORGAN

Quercus

VOYAGE
INTO THE
DEEP

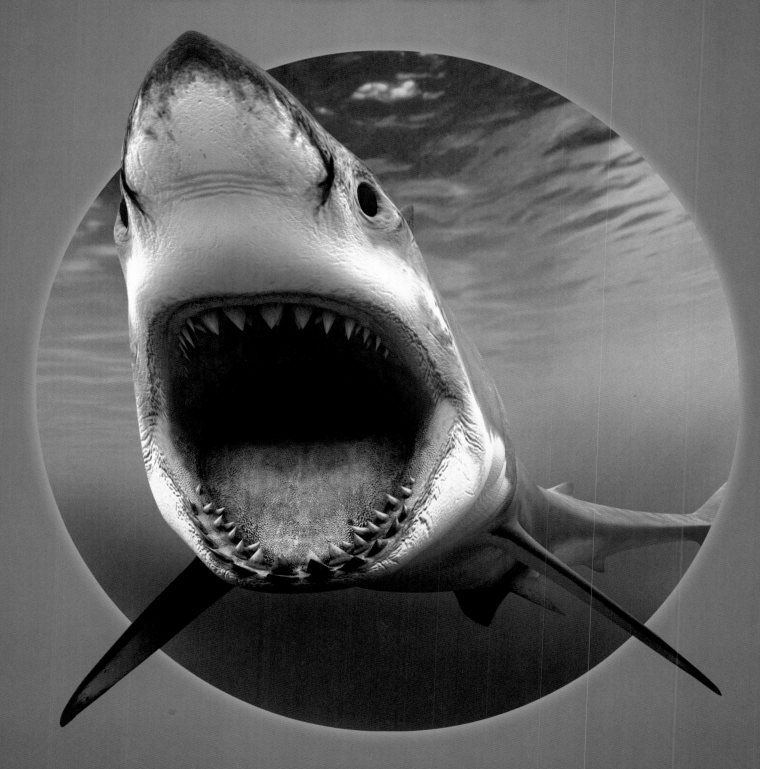

SALLY MORGAN

Quercus

CONTENTS

0211628.83.9 010273462102730106389210.948.6300011628.83.9
0273010638921O.948.6300011628.83.9 0102734621027301O6389210

PART 2 ON THE HIGH SEAS

9.6300011628.83.9
4621027301O6389210

INTRODUCTION

Get ready to launch yourself on an amazing journey to the bottom of the ocean. Your journey starts with an exploration of the fringes of the oceans – the mysterious underwater kelp forests, beautiful coral reefs and frozen polar seas – before travelling across the vast open ocean with its busy shoals of fish. Then comes the final descent, a dive into the eerie gloom of the deep, where few people have ever ventured, to explore the weird and wonderful animals that live their life in the dark.

On your journey you will see dozens of animals, from tiny jellyfish to huge whales – and not forgetting the most feared predator, the great white shark – all from the safety of your submersible. There's no need to get wet: you can see everything around you through the portholes. Hop in and close the hatch – your incredible journey is about to begin.

PORTHOLE VIEW

You can look out into the water through the portholes on the submersible. The glass in the portholes is really thick, so there is no risk of its cracking under the pressure of the deep ocean. Some animals may even come up really close to peer at you inside!

DIGITAL DISPLAY
The latest cameras zoom in on the animal, giving you a great view of its body – it's almost as if you could put your hand out and touch it.

HOW DEEP?
The green bar on the depth gauge tells you the minimum and maximum depths in the water where the animal you are looking at can be found – from the shallows to the deepest ocean floor.

DESCRIPTION The rugose squat lobster is orange-brown in colour with dark stripes on its carapace (shell) and abdomen. The squat lobster's front pair of legs are armed with long pincers. The tips of these claws are white. There are four further pairs of legs, although the back pair cannot usually be seen as they are hidden under the carapace.

EYES ON STALKS
The squat lobster has a large pair of stalked eyes, which give it good vision on the dark seabed. Between the eyes is a single long spike called a rostrum, while there is also a shorter spike to each side.

BRISTLY MOUTH
The squat lobster sweeps up food into its mouth using bristles on its mouthparts. This creature also has two pairs of antennae. The shorter pair are used for tasting and smelling, while the longer pair are used for feeling.

RUGOSE SQUAT LOBSTER

LATIN NAME *Munida rugosa*

The rugose, or long-clawed, squat lobster is one of many species of squat lobster that occur around the world. Most squat lobsters are shy and hide in crevices and under rocks during the day, but the rugose squat lobster usually sits outside its burrow or rock ledge with its claws raised in defence. It retreats into its burrow when disturbed. Usually the squat lobster walks slowly across the seabed with its abdomen tucked in, but if threatened, it uses its abdomen to swim rapidly backwards to escape.

Depth gauge:
0 m / 0 FT
100 m / 330 FT
500 m / 1640 FT
1000 m / 3280 FT
2000 m / 6560 FT
3000 m / 9840 FT
4000 m / 13120 FT
6000 m / 26240 FT
12000 m / 39360 FT

DEPTH GAUGE
3–150 m
(10–500 FT)

SIZE
10 CM
4 IN

DISTRIBUTION
The cold waters of the North Atlantic Ocean and North Sea, as far south as Madeira, and also in the Mediterranean Sea.

HABITAT This squat lobster is found on rock ledges or in burrows dug in the sand.

DIET Squat lobsters are active at night, scavenging for dead creatures on the seabed. They also catch small animals, such as krill, with their pincers.

BEHAVIOUR The squat lobster moults several times before it reaches full size. To moult, the shell splits across the top, allowing the lobster to step out of its old 'skin'. It hides in its burrow while its new soft shell becomes hard. Squat lobsters also spend a lot of time grooming themselves to remove larvae and spores.

RELATIVES The squat lobster may look like a small lobster but it is more closely related to the hermit crab. The rugose squat lobster belong to the genus *Munida*, of which there are more than 100 species distributed around the world's oceans, both in tropical and cold water.

RANGEFINDER 5 CM (2 IN)

32

33

WHERE IN THE WORLD?
Want to know where you can find this animal? The coloured areas on the world map show you its distribution. If it lives at the Poles, there are special polar maps.

PERISCOPE VIEW

You can't see which animals are swimming above the submersible through the portholes – but don't worry, you can spy on them with the periscope. This simple tube with its clever arrangement of mirrors allows you to see the animal, but it can't see you!

UP PERISCOPE
The periscope can be raised to get really close to creatures in the water without frightening them away.

HOW BIG?
Sometimes you will see only part of the animal through the periscope, so this clever piece of kit shows you the shape and maximum length of the creature to get the bigger picture.

LONGSNOUT SEAHORSE

LATIN NAME *Hippocampus reidi*

Seahorses are among the most easily recognized fish, with a head shaped like a horse's and a long body. Seahorses lie upright in the water and are propelled forwards by small fins. They live within sea grass, anchoring themselves by wrapping their tail around a blade of grass. Seahorses are protected, as their numbers have fallen steeply, mostly due to the pet trade and water pollution.

DISTRIBUTION The coastal waters of the western Atlantic Ocean, from North Carolina to Brazil.

HABITAT Seahorses like tropical water where there is coral reef, sea grass beds or mangrove swamps.

DIET They eat small animals, such as plankton and tiny fish.

DESCRIPTION The longsnout seahorse has a long body and small fins. The tail is also long and is prehensile, which means it can wrap around objects such as sea grass. The male is bright orange and the female is yellow. Both have small brown or white spots.

RELATIVES There are about 30 different types of seahorse. Seahorses are related to the pipe fish and seadragons.

BEHAVIOUR Seahorses pair for life. The female lays as many as 1000 eggs in the brood pouch of the male. The eggs hatch after about 14 days. The male releases the tiny babies at night, when they have a better chance of surviving. However, only two or three from each hatch survive to adulthood.

TOUGH ARMOUR
The body of this male is covered in protective plates that lie just beneath the skin. Males are territorial, staying within 1 sq m (10 sq ft) of their habitat, while females move over an area a hundred times that size.

SUCKING SNOUT
The seahorse is an ambush predator, lying in wait until prey passes close by. Then it lunges forwards and sucks up its victim with its long snout.

28

29

UP CLOSE
Background information is supplied for every animal, such as its Latin name, what it eats, and what relatives it has in the animal world.

ESSENTIAL INFORMATION
These little fact boxes highlight points of interest on the animal's body, such as its teeth, fins or weapons.

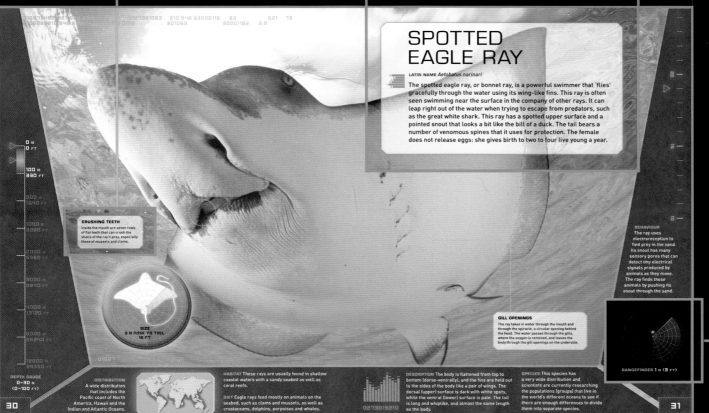

SPOTTED EAGLE RAY

LATIN NAME *Aetobatus narinari*

The spotted eagle ray, or bonnet ray, is a powerful swimmer that 'flies' gracefully through the water using its wing-like fins. This ray is often seen swimming near the surface in the company of other rays. It can leap right out of the water when trying to escape from predators, such as the great white shark. This ray has a spotted upper surface and a pointed snout that looks a bit like the bill of a duck. The tail bears a number of venomous spines that it uses for protection. The female does not release eggs: she gives birth to two to four live young a year.

CRUSHING TEETH
Inside the mouth are seven rows of flat teeth that can crush the shells of the ray's prey, especially those of mussels and clams.

BEHAVIOUR
The ray uses electroreception to find prey in the sand. Its snout has many sensory pores that can detect tiny electrical signals produced by animals as they move. The ray finds these animals by pushing its snout through the sand.

GILL OPENINGS
The ray takes in water through the mouth and through the spiracle, a circular opening behind the head. The water passes through the gills, where the oxygen is removed, and leaves the body through the gill openings on the underside.

SIZE
5 m NOSE TO TAIL
16 FT

DISTRIBUTION
A wide distribution that includes the Pacific coast of North America, Hawaii and the Indian and Atlantic Oceans.

HABITAT These rays are usually found in shallow coastal waters with a sandy seabed as well as coral reefs.

DIET Eagle rays feed mostly on animals on the seabed, such as clams and mussels, as well as crustaceans, dolphins, porpoises and whales.

DESCRIPTION The body is flattened from top to bottom (dorso-ventrally), and the fins are held out to the sides of the body like a pair of wings. The dorsal (upper) surface is dark with white spots, while the ventral (lower) surface is pale. The tail is long and whiplike, and almost the same length as the body.

SPECIES This species has a very wide distribution and scientists are currently researching the populations (groups) that live in the world's different oceans to see if there are enough differences to divide them into separate species.

30

31

TURRET VIEW

Some animals are huge, making them far too large to see through the periscope or the porthole. Pop up into the submersible's turret to get a good all-around view. Watch quietly as these giants of the ocean world swim by.

HOW CLOSE?
The submersible is fitted with the latest sonar to locate animals in the water. This screen tells you how far away it is – from a few centimetres to several metres.

RANGEFINDER 1 m (3 FT)

THE OCEANS

Water covers more than 70 per cent of the Earth's surface, so it is no surprise that the globe is often called 'the blue planet'. Plunging to depths of 3–4 km (10000–13000 ft), with even deeper ocean trenches, the world's oceans and seas hold about 97 per cent of the Earth's water.

Extending from the coasts is a zone of shallow water, known as the continental shelf. During the last ice age, this was land. Since then, the Earth has got warmer and the ice sheets have melted, causing the sea levels to rise. This shallow water (seen as a band of paler blue around the continents and islands on the photograph to the right) is home to more fish than anywhere else. The seabed then slopes steeply to the floor of the ocean. The dark blue areas on the photograph represent the deep ocean, which plunges to depths of up to 11000 m (36000 ft).

The global ocean is divided by geographers into five oceans, in descending order of size: the Pacific, Atlantic, Indian, Southern and Arctic Oceans. Large expanses of saltwater connected with an ocean are known as seas.

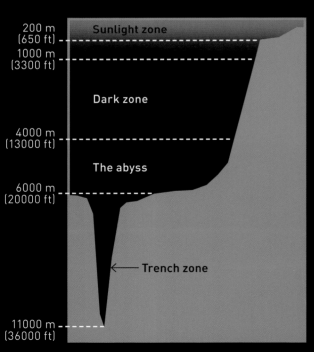

200 m (650 ft)	Sunlight zone
1000 m (3300 ft)	
	Dark zone
4000 m (13000 ft)	
	The abyss
6000 m (20000 ft)	
	← Trench zone
11000 m (36000 ft)	

OCEAN ZONES
The ocean is divided into five zones that extend from the surface to the bottom of the deepest ocean trenches. Most animals live in the sunlight and twilight zones, but a few survive in the dark zone and the even deeper abyss. Some animals can be found living on the seabed of the deepest oceans, at depths of 4000 m (13000 ft) or more.

ATLANTIC OCEAN

This long, thin ocean lies between the Americas and Europe and Africa. Down its middle runs the underwater chain of mountains called the Mid-Atlantic Ridge.

AREA 76.7 million sq km (29.6 million sq miles)

DEEPEST POINT Puerto Rico Trench: 8605 m (28232 ft)

BERING SEA

HUDSON BAY

GULF OF MEXICO

CARIBBEAN SEA

PACIFIC OCEAN

The Pacific is the world's largest ocean, covering just under 30 per cent of the world's surface – that's more than all the land mass. Its name means 'peaceful'.

AREA 155.6 million sq km (60 million sq miles)

DEEPEST POINT Mariana Trench: 10924 m (35830 ft)

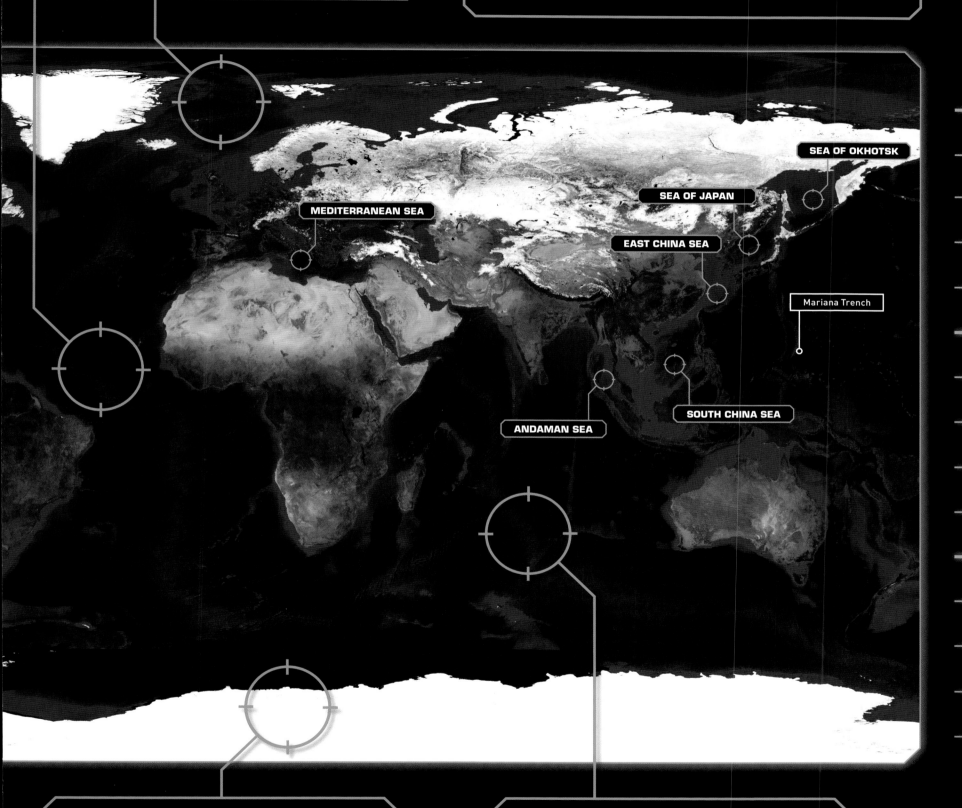

ARCTIC OCEAN

The Arctic is the smallest of the world's oceans, lying around the North Pole. In winter it is covered by sea ice, of which about half melts in summer.

AREA 14 million sq km (5.4 million sq miles)

DEEPEST POINT Eurasia Basin: 5450 m (17881 ft)

10 LARGEST SEAS OF THE WORLD

1.	South China Sea	2,974,600 sq km (1,148,495 sq miles)
2.	Caribbean Sea	2,515,900 sq km (971,390 sq miles)
3.	Mediterranean Sea	2,510,000 sq km (969,115 sq miles)
4.	Bering Sea	2,261,000 sq km (872,975 sq miles)
5.	Gulf of Mexico	1,507,600 sq km (582,085 sq miles)
6.	Sea of Okhotsk	1,392,100 sq km (537,490 sq miles)
7.	Sea of Japan	1,012,900 sq km (391,080 sq miles)
8.	Hudson Bay	730,100 sq km (281,890 sq miles)
9.	East China Sea	664,000 sq km (256,370 sq miles)
10.	Andaman Sea	564,900 sq km (218,110 sq miles)

SEA OF OKHOTSK

SEA OF JAPAN

MEDITERRANEAN SEA

EAST CHINA SEA

Mariana Trench

ANDAMAN SEA

SOUTH CHINA SEA

SOUTHERN OCEAN

This is the world's fourth largest ocean and was only recognized in 2000. It includes all the ocean lying below 60° south, much of which is frozen in winter.

AREA 20.3 million sq km (7.8 million sq miles)

DEEPEST POINT South Sandwich Trench: 7235 m (23731 ft)

INDIAN OCEAN

The Indian Ocean is bordered by Africa, Asia and Australia, while on its southern boundary its waters mix with those of the Southern Ocean.

AREA 68.6 million sq km (26.5 million sq miles)

DEEPEST POINT Java Trench: 7258 m (23812 ft)

AT THE FRINGES

Our journey begins where the water meets the land, at the fringes of the great global ocean that laps the shores of every continent and every island.

Coastlines extend for thousands upon thousands of kilometres around the world. This is an ever-changing landscape created by the waves and the tides. Along this narrow strip of land and water there are towering cliffs and rocky shores, sandy beaches, mud flats and salt marshes, and tropical mangrove swamps. Just offshore there are thick underwater kelp forests and tropical coral reefs, while in the polar regions the animals of the fringes have to cope with the extreme cold and the frozen sea.

Coasts are fun places to visit, to play on the sand and swim in the surf. It's hard to believe that life is tough for the animals living here. For part of the day the tide comes in, covering everything with salt water. When the tide retreats, the shore is left exposed to the sun, wind and rain. The animals that survive here have to be able to cope with these daily changes. On the shore, sea lions and seals may be seen basking before taking to the water to hunt. In the shallow coastal waters, crabs and sea spiders are spotted creeping along the seabed.

BENEATH THE TIDES

As we descend a few metres into the water, the sunlight piercing the surface allows us to pick out many species living in these coastal areas.

If you look up, you can see the waves rolling through the water above you, but the water below is almost still. The sunlight means that there are plenty of seaweeds, which make their food using the energy from sunshine in a process called photosynthesis. Here, the animals are never uncovered by the tides, although they live in shallow water. However, they have to cope with changing temperatures because the summer sun warms up the water. The constant churning of the water brings in plenty of food, so these coastal waters are among the most heavily populated in the world.

A great variety of animals live in these shallow waters, from turtles that pull themselves onto the beach to lay their eggs, to seastars and anemones that cling to the rocks, to weird-looking wobbegong sharks and eagle rays that lie on the sandy seabed waiting for their prey.

KELP FOREST

This underwater forest is formed from kelp, a fast-growing seaweed that is found in shallow coastal waters. The dense fronds, moving gently in the currents, form the perfect hiding place for many different fish, which in turn attract predators such as sharks.

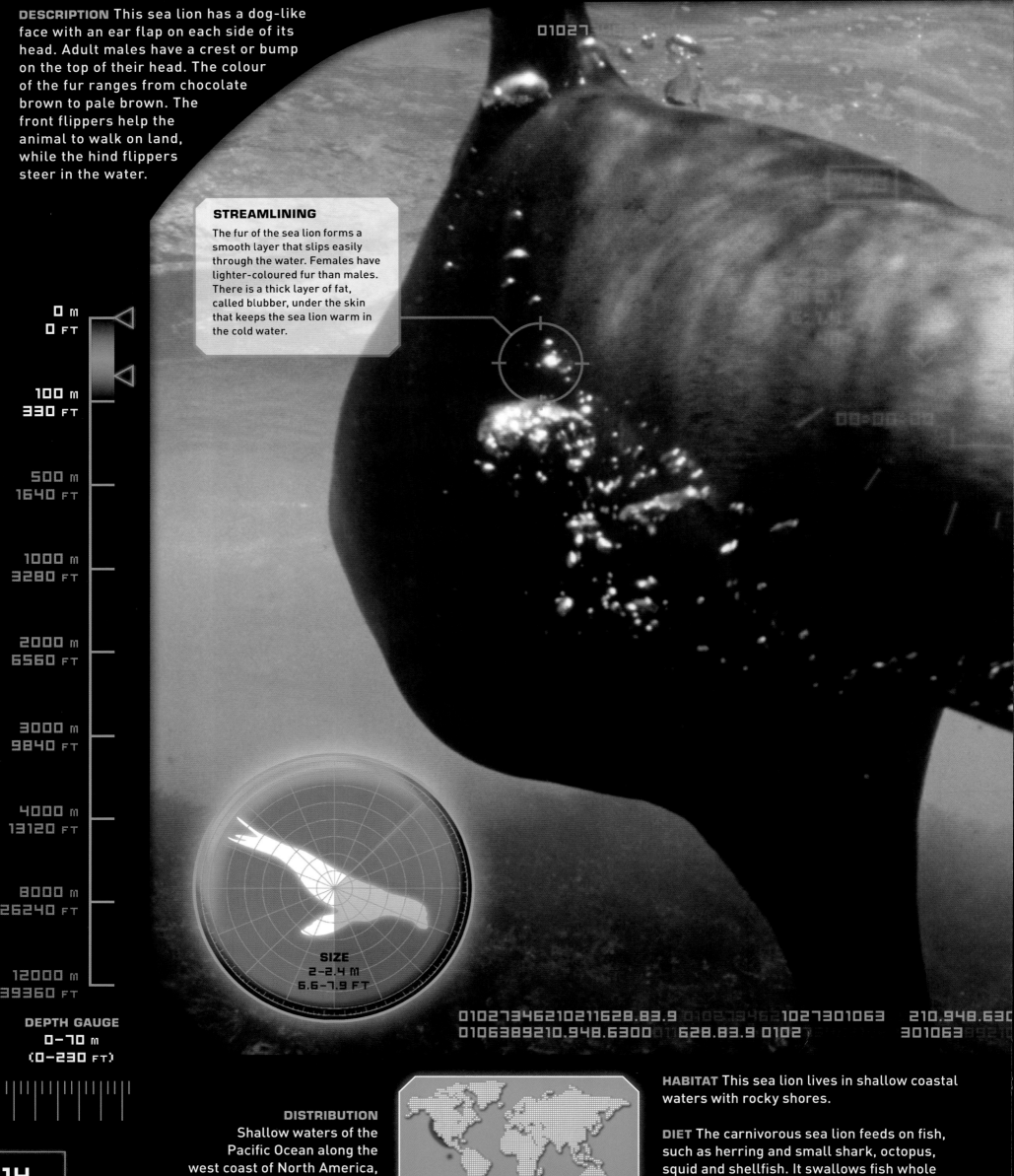

DESCRIPTION This sea lion has a dog-like face with an ear flap on each side of its head. Adult males have a crest or bump on the top of their head. The colour of the fur ranges from chocolate brown to pale brown. The front flippers help the animal to walk on land, while the hind flippers steer in the water.

STREAMLINING

The fur of the sea lion forms a smooth layer that slips easily through the water. Females have lighter-coloured fur than males. There is a thick layer of fat, called blubber, under the skin that keeps the sea lion warm in the cold water.

01027

0 m
0 FT

100 m
330 FT

500 m
1640 FT

1000 m
3280 FT

2000 m
6560 FT

3000 m
9840 FT

4000 m
13120 FT

8000 m
26240 FT

12000 m
39360 FT

DEPTH GAUGE
0–70 m
(0–230 FT)

SIZE
2–2.4 m
6.6–7.9 FT

DISTRIBUTION
Shallow waters of the Pacific Ocean along the west coast of North America, and in the Galápagos Islands.

HABITAT This sea lion lives in shallow coastal waters with rocky shores.

DIET The carnivorous sea lion feeds on fish, such as herring and small shark, octopus, squid and shellfish. It swallows fish whole and crushes shellfish.

106389210.348.630027346210273010638921D.94

CALIFORNIAN SEA LION

LATIN NAME *Zalophus californianus*

The Californian sea lion is a playful and intelligent seal. It is also very noisy, with its barking and honking – and roaring like a lion, hence its name 'sea lion'. It is the fastest swimmer of all the seals, reaching speeds of 40 kph (25 mph) when it needs to escape from a predator such as a killer whale or great white shark. These seals are social animals, living in large groups of up to 1000 individuals.

BY A WHISKER

The sea lion's stiff whiskers are very sensitive to movements in the water such as currents, so they help the sea lion find its way around in the dark. Each long whisker is called a vibrissa. In the past, people used the whiskers as pipe cleaners!

BEHAVIOUR Although sea lions spend most of their time at sea, the females have to return to land to give birth to their pup. These mammals hunt at sea for several days, diving under the water for three to four minutes, although they can stay under for up to 40 minutes. When diving, sea lions close their nostrils.

RELATIVES The sea lion belongs to the family of eared seals, within which there are 16 species of sea lion and fur seal, including the Steller sea lion and northern fur seal. Many fur seals were hunted for their fur, but hunting bans have allowed them to recover.

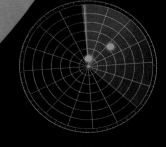

RANGEFINDER 50 cm (20 in)

GREEN TURTLE

LATIN NAME *Chelonia mydas*

The green turtle gets its name from its greenish flesh. It is a large sea turtle, weighing between 150 and 230 kg (330–500 lb). Turtles are powerful swimmers and spend most of their life at sea, but unlike fish, they have lungs and come to the surface to breathe. The female green turtle returns to the sandy beach where she was born to lay her eggs, often travelling for thousands of kilometres across the ocean.

PROTECTIVE SHELL

The hard shell, or carapace, protects the turtle from predators. It is covered by scale-like structures called scutes. Unlike the tortoise, the turtle cannot pull its head inside its shell.

0 m
0 FT

100 m
330 FT

500 m
1640 FT

1000 m
3280 FT

2000 m
6560 FT

3000 m
9840 FT

4000 m
13120 FT

8000 m
26240 FT

12000 m
39360 FT

DEPTH GAUGE
0–100 m
(0–330 FT)

DISTRIBUTION Tropical and subtropical oceans around the world.

HABITAT Young turtles spend their first few years in the open ocean. When they are mature, they return to shallow water where there are coral reefs and sea grass beds.

DIET Young turtles are carnivores, feeding on shrimp, jellyfish and corals. Adult turtles are herbivores and eat mostly algae and sea grass.

DESCRIPTION This turtle has an oval shell in shades of brown, green and black. The underside, called the plastron, is white or yellow.

SUBSPECIES There is one species of green turtle, but there are two main groups, or populations, which are found in different parts of the world: the Atlantic and the Pacific green turtles. Both groups are endangered.

BEHAVIOUR The baby turtles hatch together and all help to dig their way out of the sand, emerging at night. They find their way to the water by looking for the bright horizon of the ocean, but they can be confused by bright lights from buildings. Many are eaten by predators such as birds and lizards. Only one or two from each nest will survive to adulthood.

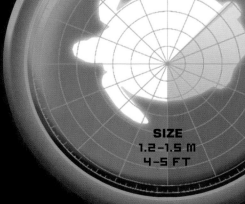

SIZE
1.2–1.5 m
4–5 FT

POWERFUL FLIPPERS

The green turtle has paddle-like flippers instead of legs, which propel it through the water at speeds of up to 56 kph (35 mph). The female also uses her flippers to dig a hole in the sand to bury her eggs.

01027346210211628.83.9   01027346 1027301063 210.948.63000116   .83
0106389210.948.6300   628.83.9 0102   34621027   301063   30001162   .3.9

THUMPING SOUNDS

The male grinds teeth in his throat to produce a thumping sound that can be heard by divers. This loud noise attracts females and warns other males to stay away from nesting or feeding sites.

SEX CHANGE

Amazingly, these fish can actually change their sex. A sex change happens when there are too many males or females, so a few individuals change their sex to even up the numbers. Some individuals may change sex many times in their lifetime.

0 m
0 FT

100 m
330 FT

500 m
1640 FT

1000 m
3280 FT

2000 m
6560 FT

3000 m
9840 FT

4000 m
13120 FT

8000 m
26240 FT

12000 m
39360 FT

DEPTH GAUGE
0–30 m
(0–100 FT)

01027   46210   628.83.9 010   3462102730106

DISTRIBUTION
The eastern central Pacific Ocean, along the Californian coast from Monterey Bay to Baja California.

HABITAT These fish are found in shallow, rocky reefs where there are clear water and kelp, as well as crevices and small caves in which to hide.

DIET The garibaldi is a carnivore, feeding on small invertebrates on the seabed such as sponges, and plankton floating in the water.

GARIBALDI

LATIN NAME *Hypsypops rubicundus*

These colourful little fish may look cute and friendly, but the males are highly aggressive. They defend a small territory and chase off other fish that come too close. They even attack divers! Their unusual name, garibaldi, comes from the 19th-century Italian revolutionary Giuseppe Garibaldi, whose soldiers wore red jackets. Most garibaldi are found along the coast of California.

SIZE
28–34 CM
10–13 IN

BEHAVIOUR

The male is the nest-builder, and he spends several weeks preparing a special mat of algae on the seabed. Once it is ready, he does a dance, dipping his head up and down, to attract females. The females inspect the nests, and choose the males with the best ones.

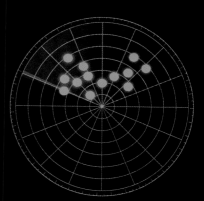

RANGEFINDER 1 m (3 FT)

DESCRIPTION Adults are bright orange with a deep body and a long, tapering dorsal fin. The young fish, known as juveniles, are orange with iridescent blue spots on their fins and body. Their fins are also outlined in blue. These blue markings fade by the time the fish reach five years old.

RELATIVES The garibaldi is the largest member of the damselfish family, of which there are more than 200 species worldwide. These are fish that defend territories and build nests for their eggs. Most damselfish are brightly coloured and live in saltwater, although some species survive in freshwater.

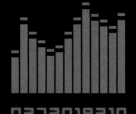
0273019210

EAFY
EADRAGON

NAME *Phycodurus eques*

st glance, this fish looks like a piece of seaweed floating
e water. This perfect camouflage is helped by the
that the fish remains motionless for days at a time.
rtunately, the leafy seadragon is threatened by
ectors who sell it to the pet trade. It is also sensitive
ollution in the water, especially sewage.

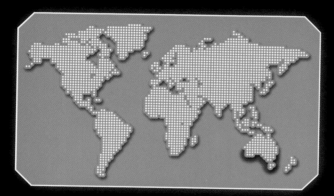

DISTRIBUTION Found only in the
shallow coastal waters of southern and
western Australia.

HABITAT These fish are found in shallow
water where there are rocky reefs with kelp
beds, and in sea grass meadows with patches of
bare sand.

DIET Leafy seadragons feed on plankton,
small fish and small crustaceans such
as shrimp.

DESCRIPTION The fish has a long, thin body
covered in armoured plates with spines for
protection. The skin produces outgrowths that
look like pieces of seaweed, which provide
excellent camouflage.

RELATIVES The leafy seadragon
is closely related to the weedy
seadragon, which has small
leaf-like outgrowths.

BEHAVIOUR The female leafy
seadragon lays up to 250 pink eggs
on the brood patch of the male.
This is a special patch of skin on
the underside of his tail. The eggs
hatch after about 60 days, and the male
releases the tiny fish by shaking his tail

MOVING ALONG

The leafy seadragon swims slowly
through the water by moving its
transparent pectoral and dorsal fins.
Although it often remains motionless
for long periods, this fish can travel
about 100 m (330 ft) in an hour.

01027 3452 0211628 83.9 0102 346207010

SIZE
35-45 CM
14-18 IN

LUGE

n

FT)

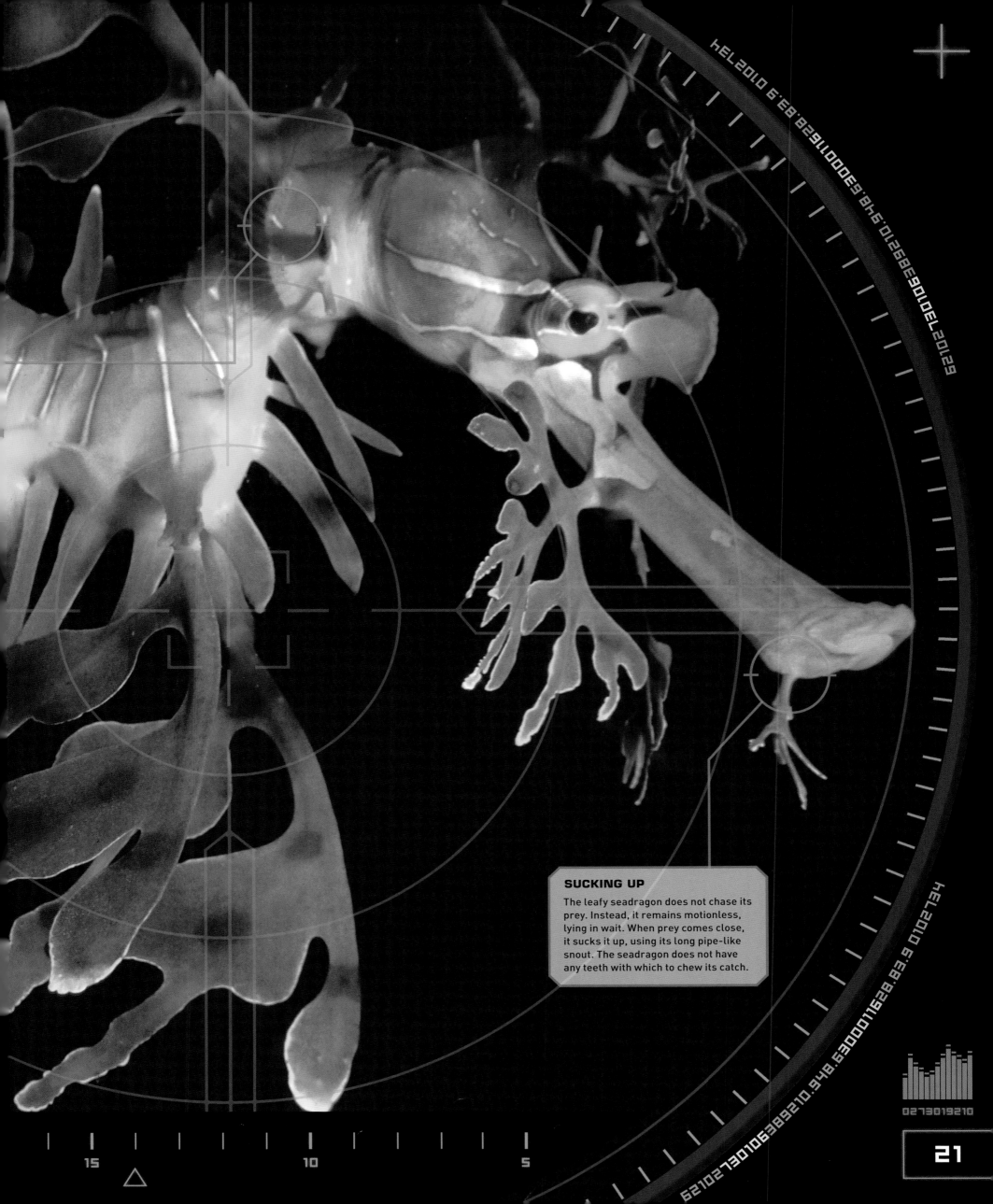

SUCKING UP

The leafy seadragon does not chase its prey. Instead, it remains motionless, lying in wait. When prey comes close, it sucks it up, using its long pipe-like snout. The seadragon does not have any teeth with which to chew its catch.

DESCRIPTION Cuttlefish are related to octopus and squid. They are all cephalopods, which are animals with a head and tentacles. The pharaoh has eight short arms and two longer arms. It has a short, broad body with lateral fins down the sides. This cuttlefish weighs up to 5 kg (11 lb).

SIZE
45–60 CM
18–24 IN

0 m
0 FT

100 m
330 FT

500 m
1640 FT

1000 m
3280 FT

2000 m
6560 FT

3000 m
9840 FT

4000 m
13120 FT

8000 m
26240 FT

12000 m
39360 FT

GRIPPING TENTACLES
Each long tentacle ends in a pad with 200 tiny suction discs. The cuttlefish shoots out these tentacles to grab its prey, while the shorter tentacles, also with suckers, help to grip and push the prey into the mouth.

DEPTH GAUGE
0–200 m
(0–660 FT)

DISTRIBUTION
From East Africa to Japan and Australia, including in the Red Sea and the Persian Gulf.

HABITAT Pharaoh cuttlefish need shallow tropical water where there is coral reef or sandy or muddy seabeds.

DIET These predatory molluscs feed on fish and invertebrates, such as shrimp and crab.

PHARAOH CUTTLEFISH

LATIN NAME *Sepia pharaonis*

The pharaoh cuttlefish is coloured bright red and orange while sporting an iridescent band of blue along the base of its fins. Cuttlefish can change the colour of their skin almost instantly, with waves of colours spreading over the surface. The colours are signals, some to attract a mate, while others are warnings. The colours can also provide camouflage when the cuttlefish rests on the seabed.

LATERAL FINS

The cuttlefish glides slowly through the water, moving its lateral fins in a wave-like motion. This creature varies its depth using the shell inside its body. It adds gas to the shell to move up and removes gas to sink.

BEHAVIOUR The female pharaoh cuttlefish lays as many as 1500 eggs which are about the size of golf balls, positioning them in crevices and under overhangs in rocks. The eggs are left to develop and hatch alone. Two weeks later, miniature cuttlefish emerge. The female dies soon after spawning.

SPECIES Until now, biologists have placed all the pharaoh cuttlefish in one species. However, recent research suggests that there may actually be five species, one in each of the different parts of the pharaoh's range, such as the Red Sea, Persian Gulf and western Pacific.

RANGEFINDER 10 cm (4 in)

RED-TILED SEASTAR

LATIN NAME *Fromia monilis*

This striking seastar, with its eye-catching pattern of reds, orange and cream, is found in tropical waters. It has several different names, including marbled starfish and necklace seastar. This is an echinoderm, an animal which has a spiny body that is symmetrical around a central point. An external skeleton, formed from plates of calcium carbonate, lies just beneath the skin.

0 m
0 FT

100 m
330 FT

500 m
1640 FT

1000 m
3280 FT

2000 m
6560 FT

3000 m
9840 FT

4000 m
13120 FT

8000 m
26240 FT

12000 m
39360 FT

DEPTH GAUGE
0-25 m
(0-80 FT)

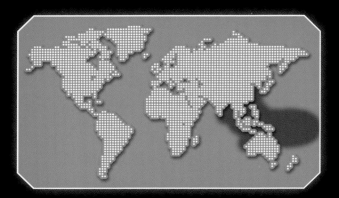

DISTRIBUTION The western Pacific and Indian Oceans.

HABITAT Seastars like a sandy seabed, but they can also be found on coral reefs and rocks.

DIET Seastars are predators, feeding on molluscs, such as clams and mussels, sea urchins and other starfish.

DESCRIPTION The red-tiled seastar is identified by large plates running down the edge of each arm, with smaller plates down the middle. The tips of the arms and the central disc have red plates.

RELATIVES The genus *Fromia* contains about 12 species, all of which are tropical starfish with five arms.

BEHAVIOUR The seastar uses its tube feet to attack prey such as bivalves (molluscs with two-part shells), using the suckers on each foot to pull the shells apart. Then it pushes its stomach out through its mouth and pours digestive juices directly into the bivalve. By feeding in this way, the seastar can prey on animals much larger than itself.

REGENERATION
If a predator pulls off one of the five arms, the seastar regrows a new one. Some species of seastar can even regrow themselves from just one arm attached to the central disc.

SIZE
28-32 CM
11-12.5 IN

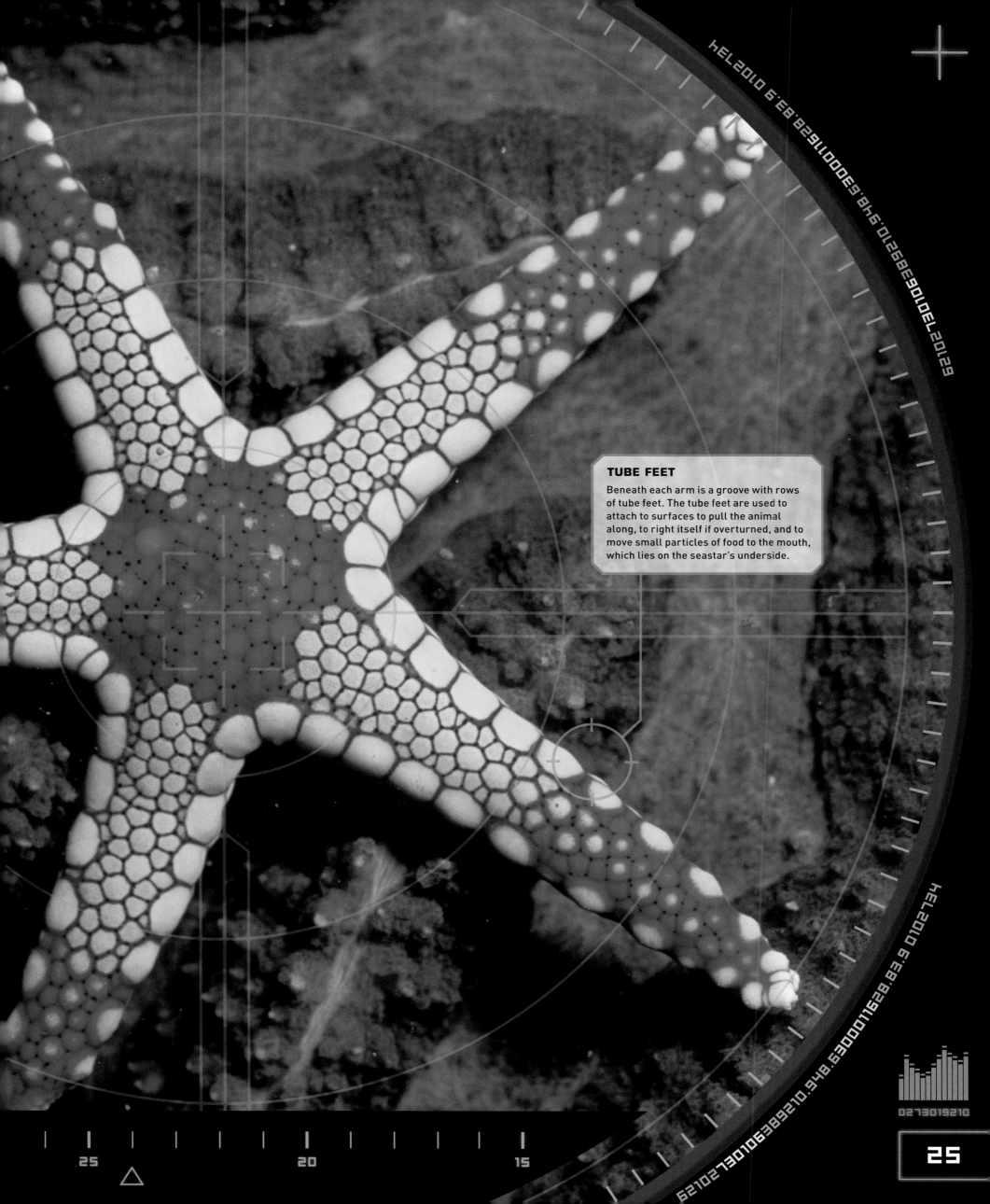

TUBE FEET

Beneath each arm is a groove with rows of tube feet. The tube feet are used to attach to surfaces to pull the animal along, to right itself if overturned, and to move small particles of food to the mouth, which lies on the seastar's underside.

EWEL ANEMONE

NAME *Corynactis* sp.

ewel anemone gets its name from its spectacular rainbow
urs of orange, red pink, electric blue, emerald green and
e. In places where there is plenty of food, the jewel
none is found in large numbers, densely packed together.
imple body consists of a tube that is closed at one end
open at the other. The open end forms the mouth,
h is surrounded by rings of tentacles.

DISTRIBUTION A wide distribution,
which includes the Atlantic, Pacific and
Indian Oceans, and the Mediterranean
and Red Seas.

HABITAT The anemone is found on the lower
shore and in shallow water underneath rocky
overhangs and on vertical rock faces.

DIET The anemone's diet includes small fish,
larval crustaceans and other food particles
that drift by.

DESCRIPTION The jewel anemone comes in
a range of colours and colour combinations,
such as green with pink tips to the tentacles.
There are 100 tentacles around the mouth,
arranged in three rings. Each tentacle ends
in a distinctive knob. A tentacle's base is
about 1 cm (0.4 in) across.

SPECIES The name jewel anemone
does not apply to a single species of
anemone, but to a group of colourful
anemones of a similar appearance.

BEHAVIOUR The jewel anemone
can reproduce asexually (needing
only one parent) by simply dividing
into two, producing two identical
individuals. This form of reproduction
enables numbers to increase quickly,
producing a colony of identically
coloured jewel anemones.

UGE
m
FT)

SIZE
15 MM
0.6 IN

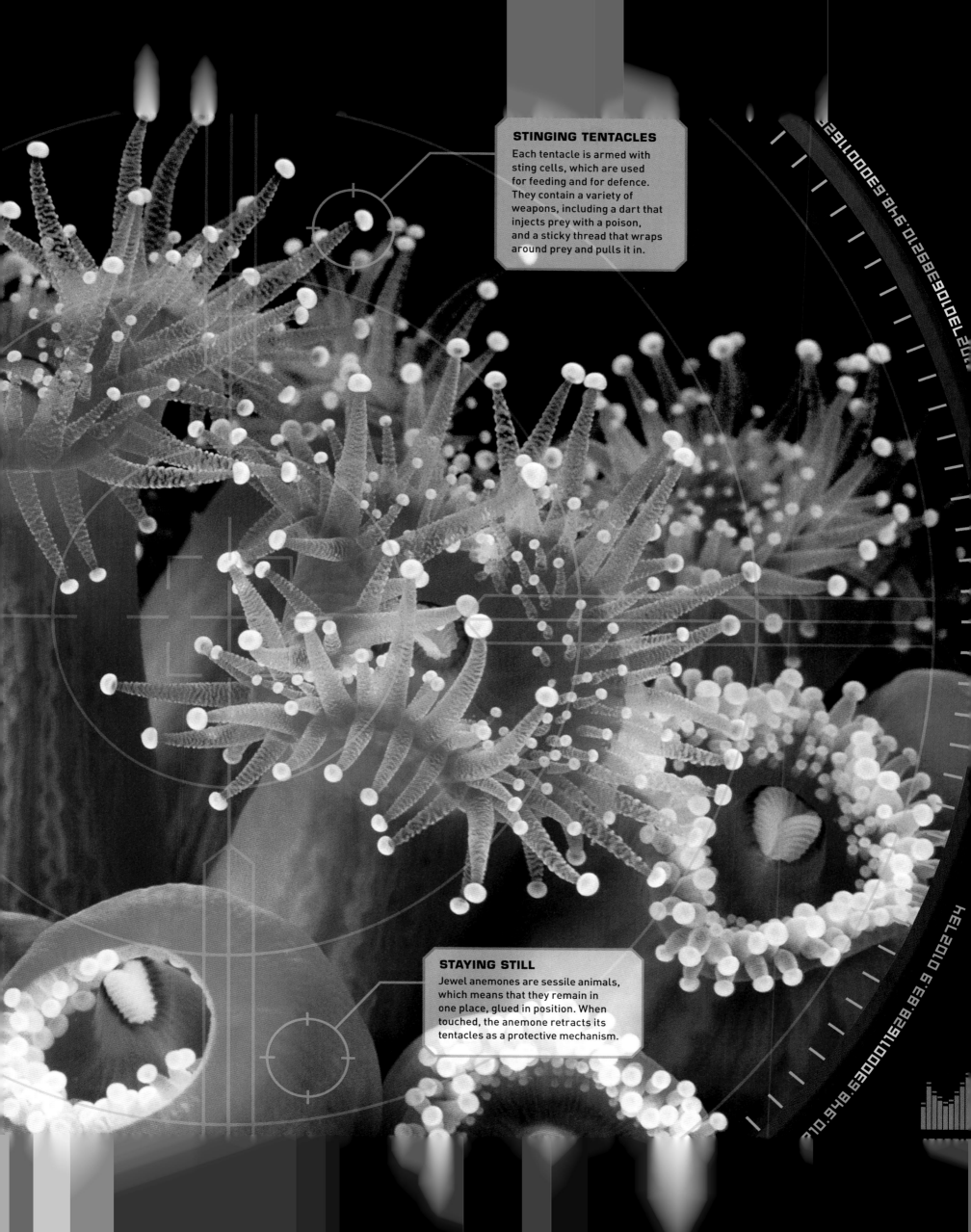

STINGING TENTACLES

Each tentacle is armed with sting cells, which are used for feeding and for defence. They contain a variety of weapons, including a dart that injects prey with a poison, and a sticky thread that wraps around prey and pulls it in.

STAYING STILL

Jewel anemones are sessile animals, which means that they remain in one place, glued in position. When touched, the anemone retracts its tentacles as a protective mechanism.

LONGSNOUT SEAHORSE

LATIN NAME *Hippocampus reidi*

Seahorses are among the most easily recognized fish, with a head shaped like a horse's and a long body. Seahorses lie upright in the water and are propelled forwards by small fins. They live within sea grass, anchoring themselves by wrapping their tail around a blade of grass. Seahorses are protected, as their numbers have fallen steeply, mostly due to the pet trade and water pollution.

0 m
0 FT

100 m
330 FT

500 m
1640 FT

1000 m
3280 FT

2000 m
6560 FT

3000 m
9840 FT

4000 m
13120 FT

8000 m
26240 FT

12000 m
39360 FT

DEPTH GAUGE
0-50 m
(0-165 FT)

DISTRIBUTION The coastal waters of the western Atlantic Ocean, from North Carolina to Brazil.

HABITAT Seahorses like tropical water where there is coral reef, sea grass beds or mangrove swamps.

DIET They eat small animals, such as plankton and tiny fish.

DESCRIPTION The longsnout seahorse has a long body and small fins. The tail is also long and is prehensile, which means it can wrap around objects such as sea grass. The male is bright orange and the female is yellow. Both have small brown or white spots.

RELATIVES There are about 30 different types of seahorse. Seahorses are related to the pipe fish and seadragons.

BEHAVIOUR Seahorses pair for life. The female lays as many as 1000 eggs in the brood pouch of the male. The eggs hatch after about 14 days. The male releases the tiny babies at night, when they have a better chance of surviving. However, only two or three from each hatch survive to adulthood.

TOUGH ARMOUR

The body of this male is covered in protective plates that lie just beneath the skin. Males are territorial, staying in just 1 sq m (10 sq ft) of their habitat, while females move over an area a hundred times that size.

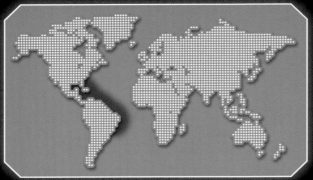

SIZE
15 CM
6 IN

SUCKING SNOUT

The seahorse is an ambush predator, lying in wait until prey passes close by. Then it lunges forwards and sucks up its victim with its long snout.

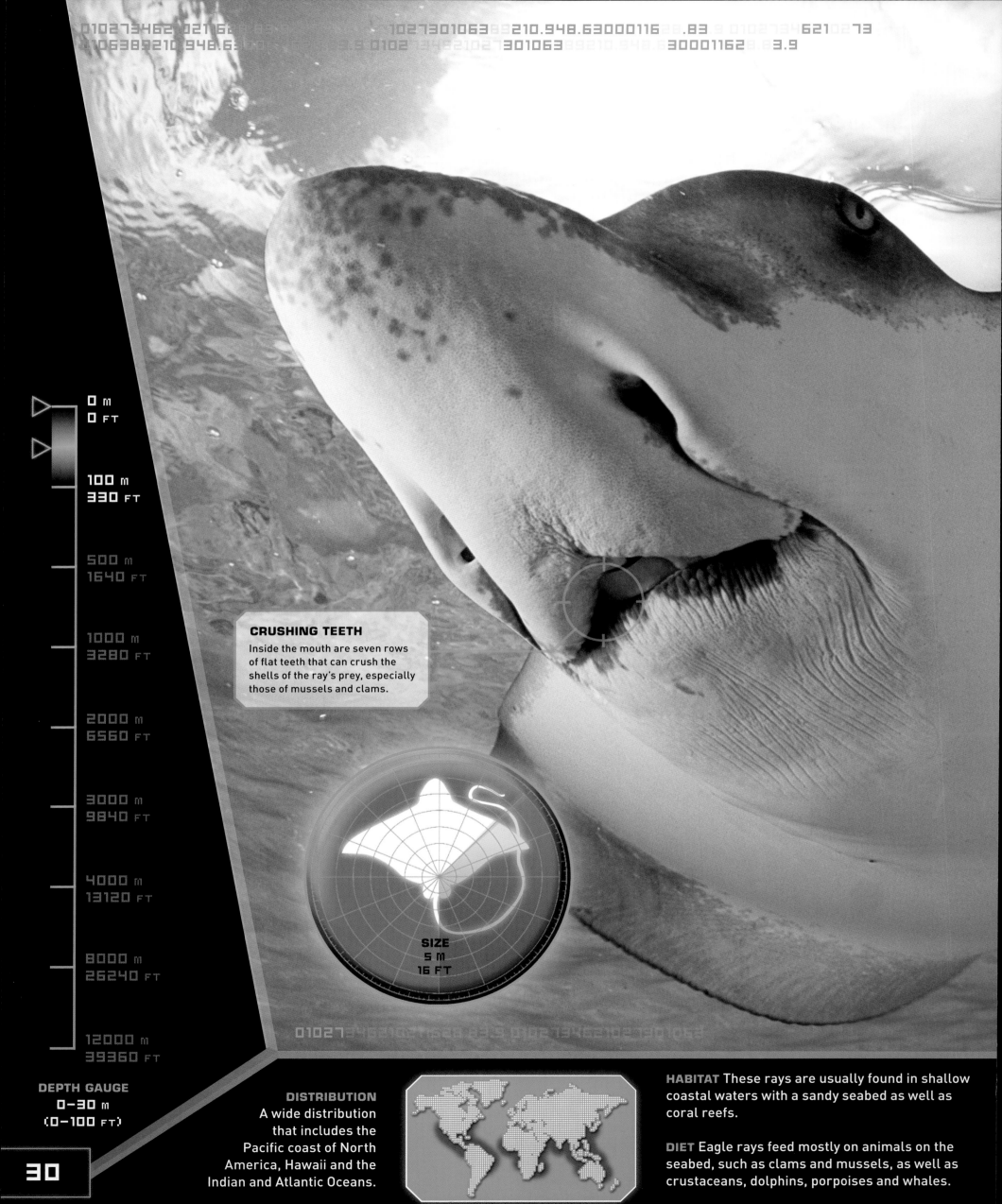

010273462 021 6 1027301063 9210.948.63000116 8.83 0102794 6210273
4106389210 948.6 9.8 0102 73462102 301063 9210.948. 30001162 8 3.9

0 m
0 FT

100 m
330 FT

500 m
1640 FT

1000 m
3280 FT

2000 m
6560 FT

3000 m
9840 FT

4000 m
13120 FT

8000 m
26240 FT

12000 m
39360 FT

DEPTH GAUGE
0–30 m
(0–100 FT)

CRUSHING TEETH
Inside the mouth are seven rows
of flat teeth that can crush the
shells of the ray's prey, especially
those of mussels and clams.

SIZE
5 M
16 FT

01027 4 21 1 8 8 9.0102 73462102 301063

DISTRIBUTION
A wide distribution
that includes the
Pacific coast of North
America, Hawaii and the
Indian and Atlantic Oceans.

HABITAT These rays are usually found in shallow
coastal waters with a sandy seabed as well as
coral reefs.

DIET Eagle rays feed mostly on animals on the
seabed, such as clams and mussels, as well as
crustaceans, dolphins, porpoises and whales.

SPOTTED EAGLE RAY

LATIN NAME *Aetobatus narinari*

The spotted eagle ray, or bonnet ray, is a powerful swimmer that 'flies' gracefully through the water using its wing-like fins. This ray is often seen swimming near the surface in the company of other rays. It can leap right out of the water when trying to escape from predators, such as the great white shark. This ray has a spotted upper surface and a pointed snout that looks a bit like the bill of a duck. The tail bears a number of venomous spines that it uses for protection. The female does not release eggs: she gives birth to two to four live young a year.

BEHAVIOUR
The ray uses electroreception to find prey in the sand. Its snout has many sensory pores that can detect tiny electrical signals produced by animals as they move. The ray finds these animals by pushing its snout through the sand.

GILL OPENINGS
The ray takes in water through the mouth and through the spiracle, a circular opening behind the head. The water passes through the gills, where the oxygen is removed, and leaves the body through the gill openings on the underside.

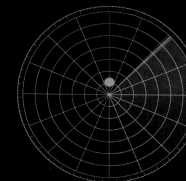

RANGEFINDER 1 m (3 ft)

DESCRIPTION The body is flattened from top to bottom (dorso-ventrally), and the fins are held out to the sides of the body like a pair of wings. The dorsal (upper) surface is dark with white spots, while the ventral (lower) surface is pale. The tail is long and whiplike, and almost the same length as the body.

SPECIES This species has a very wide distribution and scientists are currently researching the populations (groups) that live in the world's different oceans to see if there are enough differences to divide them into separate species.

0273019210

31

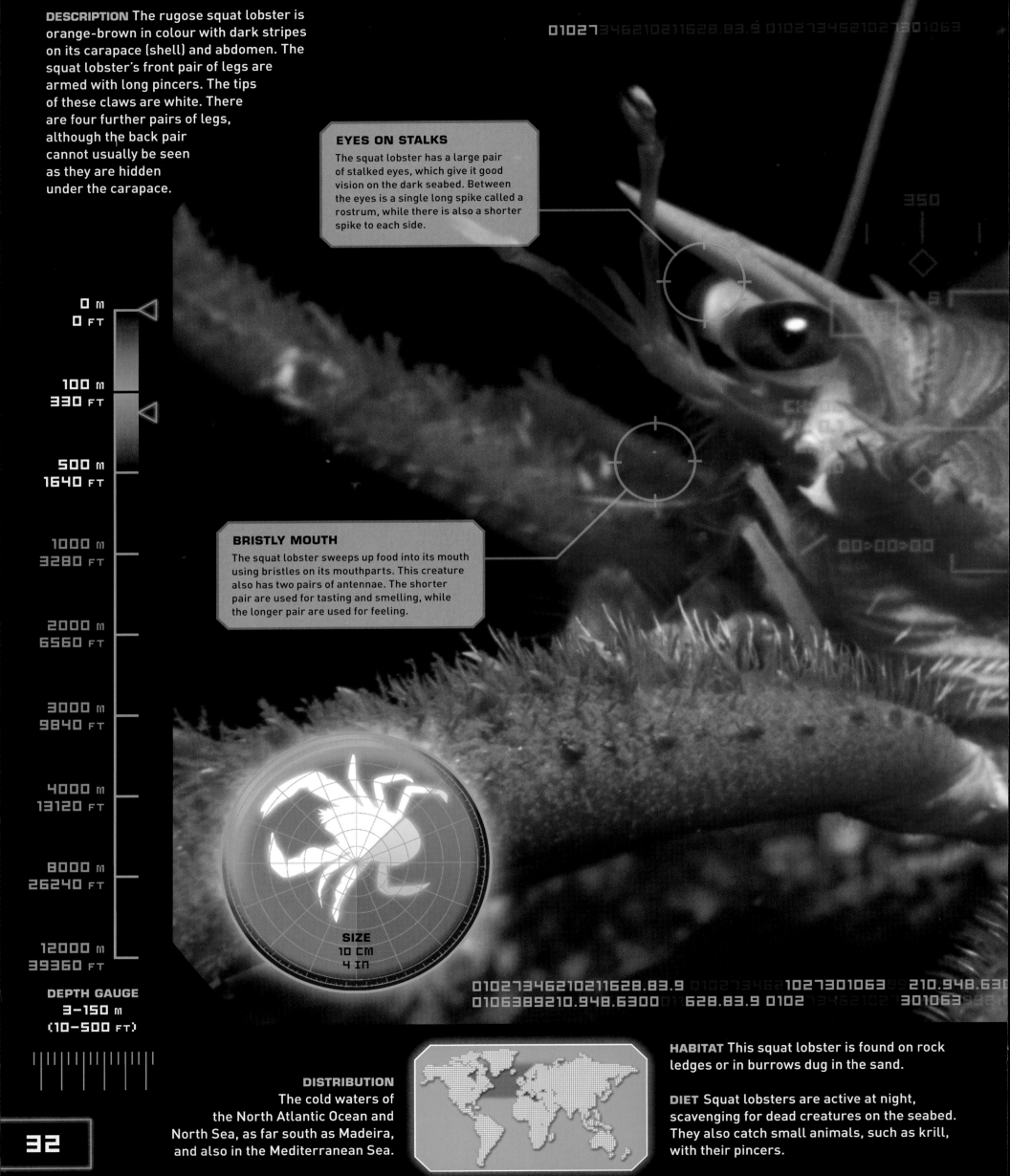

DESCRIPTION The rugose squat lobster is orange-brown in colour with dark stripes on its carapace (shell) and abdomen. The squat lobster's front pair of legs are armed with long pincers. The tips of these claws are white. There are four further pairs of legs, although the back pair cannot usually be seen as they are hidden under the carapace.

EYES ON STALKS
The squat lobster has a large pair of stalked eyes, which give it good vision on the dark seabed. Between the eyes is a single long spike called a rostrum, while there is also a shorter spike to each side.

BRISTLY MOUTH
The squat lobster sweeps up food into its mouth using bristles on its mouthparts. This creature also has two pairs of antennae. The shorter pair are used for tasting and smelling, while the longer pair are used for feeling.

0 m
0 FT

100 m
330 FT

500 m
1640 FT

1000 m
3280 FT

2000 m
6560 FT

3000 m
9840 FT

4000 m
13120 FT

8000 m
26240 FT

12000 m
39360 FT

DEPTH GAUGE
3–150 m
(10–500 FT)

SIZE
10 CM
4 IN

DISTRIBUTION
The cold waters of the North Atlantic Ocean and North Sea, as far south as Madeira, and also in the Mediterranean Sea.

HABITAT This squat lobster is found on rock ledges or in burrows dug in the sand.

DIET Squat lobsters are active at night, scavenging for dead creatures on the seabed. They also catch small animals, such as krill, with their pincers.

RUGOSE SQUAT LOBSTER

LATIN NAME *Munida rugosa*

The rugose, or long-clawed, squat lobster is one of many species of squat lobster that occur around the world. Most squat lobsters are shy and hide in crevices and under rocks during the day, but the rugose squat lobster usually sits outside its burrow or rock ledge with its claws raised in defence. It retreats into its burrow when disturbed. Usually the squat lobster walks slowly across the seabed with its abdomen tucked in, but if threatened, it uses its abdomen to swim rapidly backwards to escape.

BEHAVIOUR The squat lobster moults several times before it reaches full size. To moult, the shell splits across the top, allowing the lobster to step out of its old 'skin'. It hides in its burrow while its new soft shell becomes hard. Squat lobsters also spend a lot of time grooming themselves to remove larvae and spores.

RELATIVES The squat lobster may look like a small lobster but it is more closely related to the hermit crab. The rugose squat lobster belong to the genus *Munida*, of which there are more than 100 species distributed around the world's oceans, both in tropical and cold water.

ACY
CORPIONFISH

NAME *Rhinopias aphanes*

strange-looking shape is a lacy scorpionfish resting on the
ed. The maze-like pattern of colours helps to disrupt
ish's outline, making it difficult to spot. Also known as
let's scorpionfish, this is a sedentary fish, resting on
sea floor for much of the time. When it does move, it
ks', using its pectoral (side) fins, rather than swims.

DISTRIBUTION The western Pacific Ocean,
especially around Papua New Guinea
and off northeast Australia.

HABITAT This fish likes coral reefs, but it is
also found on rocky and sandy seabeds.

DIET It feeds on a range of animals, such as
crustaceans, squid, octopus and fish.

DESCRIPTION The scorpionfish has an
upturned mouth, with many tentacles
around the nose, and long skin flaps forming
tassels and warts on the body. The fish's
colour is variable, the background colour
ranging from black and green to yellow
and brown, with a maze-like pattern of a
contrasting colour.

RELATIVES There are eight species of
Rhinopias scorpionfish, which are found
mostly around the Pacific Ocean. All
species have excellent camouflage, but
little more is known about them
as they are rarely seen.

BEHAVIOUR Scorpionfish live on or
near the seabed, hiding in crevices
and caves. When at rest, a scorpionfish
rocks gently and quivers its fins to
create the impression of algae moving
in the current.

UGE
m
T)

SIZE
30 CM
12 IN

GAPING MOUTH

The scorpionfish has a large, wide mouth, enabling it to tackle quite big prey. Its camouflage allows it to lie in wait for its victims. The attack is over in a split second, as the unlucky passing fish is sucked into the scorpionfish's mouth.

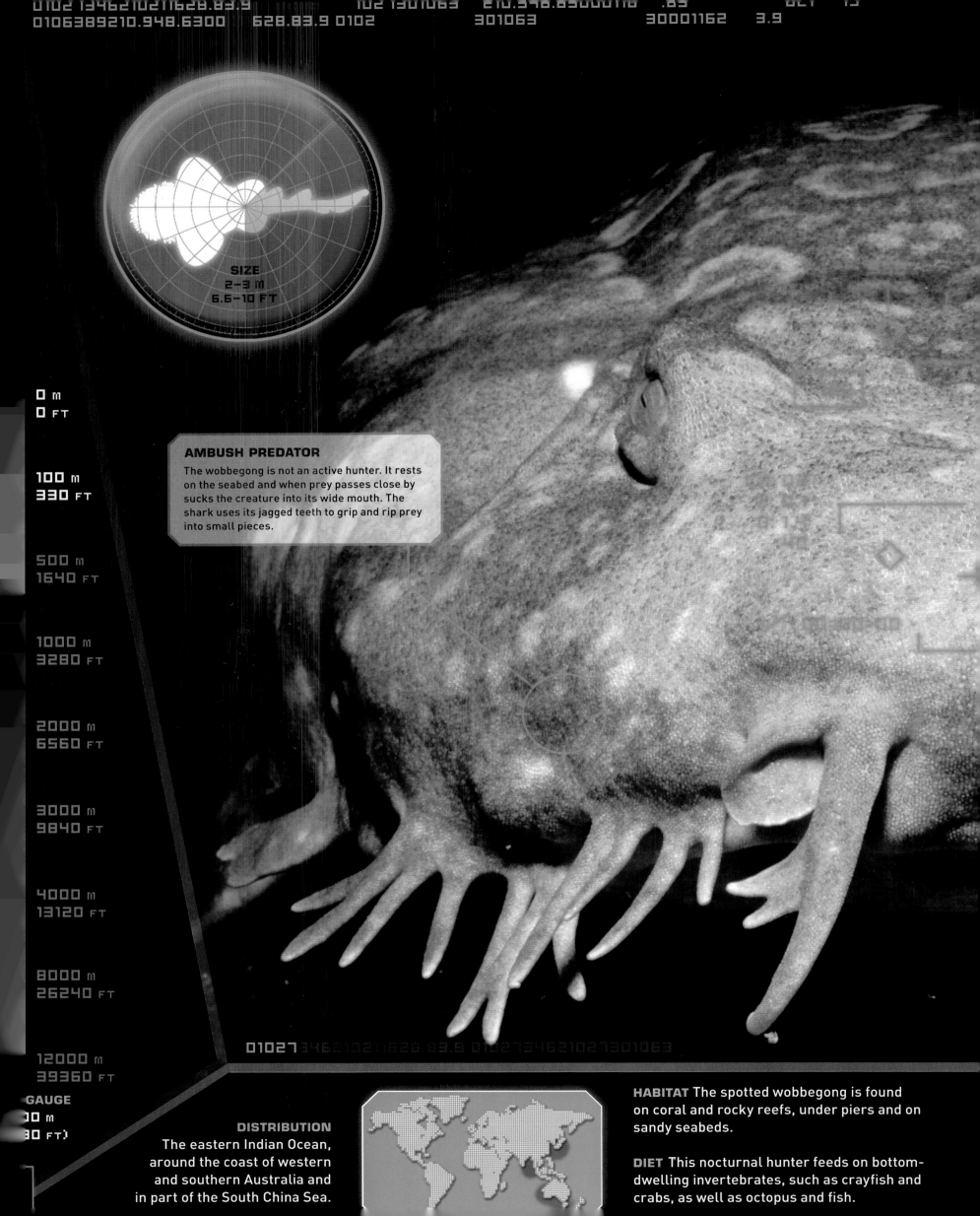

SIZE
2–3 M
6.6–10 FT

0 m
0 FT

AMBUSH PREDATOR
The wobbegong is not an active hunter. It rests
on the seabed and when prey passes close by
sucks the creature into its wide mouth. The
shark uses its jagged teeth to grip and rip prey
into small pieces.

100 m
330 FT

500 m
1640 FT

1000 m
3280 FT

2000 m
6560 FT

3000 m
9840 FT

4000 m
13120 FT

8000 m
26240 FT

01027 346210211628.83.9 0127346831027301063 .63

12000 m
39360 FT

GAUGE
30 m
30 FT)

DISTRIBUTION
The eastern Indian Ocean,
around the coast of western
and southern Australia and
in part of the South China Sea.

HABITAT The spotted wobbegong is found
on coral and rocky reefs, under piers and on
sandy seabeds.

DIET This nocturnal hunter feeds on bottom-
dwelling invertebrates, such as crayfish and
crabs, as well as octopus and fish.

SPOTTED WOBBEGONG

LATIN NAME *Orectolobus maculatus*

The wobbegong looks more like a seaweed-covered rock than a shark. It gets its name from an Aboriginal Australian word that means 'shaggy beard', which refers to the long skin flaps around its head. Wobbegongs are not aggressive sharks, but their camouflage makes them difficult to spot in the shallows, so people have trodden on them and been bitten. The wobbegong is under threat, mostly because of overfishing.

BEHAVIOUR
The female wobbegong does not lay eggs like most sharks. Instead the eggs are held inside her body and she gives birth to live young. There are up to 20 or more young, each about 20 cm (8 in) long.

FRINGE
The characteristic seaweed-like fringe is formed from skin flaps that hang around the head from the snout to just in front of the gills. This 'beard' seems to make the wobbegong melt into the seabed, making it doubly hard to spot.

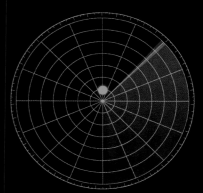

RANGEFINDER 1 m (3 FT)

DESCRIPTION This shark has a flattened body with a broad head and blunt snout. The seaweed-like skin flaps on the head create a fringe. The wobbegong's yellowish-brown back is marked with darker patches and covered with lighter-coloured rings. This coloration acts as camouflage on the seabed.

RELATIVES There are 11 species of wobbegongs in the family Orectolobidae, including the tasselled wobbegong and the ornate wobbegong. All these sharks are also called carpet sharks because of their bold skin markings. They are found in both tropical and temperate waters.

0273019210

RED-LIPPED BATFISH

LATIN NAME *Ogcocephalus darwini*

This odd-looking fish gets its name from the flattened shape of its body, which looks a bit like a bat. It is a species of anglerfish, a group of fish that use a 'fishing rod' lure to attract prey. The adults are found on the seabed but the larvae are found floating in mid-water.

0 m
0 FT

100 m
330 FT

500 m
1640 FT

1000 m
3280 FT

2000 m
6560 FT

3000 m
9840 FT

4000 m
13120 FT

8000 m
26240 FT

12000 m
39360 FT

DEPTH GAUGE
10–120 m
(30–400 FT)

DISTRIBUTION The waters around the Galápagos Islands in the Pacific Ocean.

HABITAT Red-lipped batfish are found on the seabed in shallow coastal water. Other species of batfish are found in much deeper water.

DIET This predator eats small fish, molluscs, shrimps and other crustaceans.

DESCRIPTION The body of the red-lipped batfish is flattened from top to bottom, and its pectoral and pelvic fins stick out a bit like legs, propping the fish up when on the seabed. It has distinctive red lips and a horn between the eyes. Its skin is rough like sandpaper.

RELATIVES There are about 68 species of batfish, which belong to the anglerfish order. The red-lipped batfish does not occur as deep as some of its close relatives, which are found to depths of 4000 m (13000 ft).

BEHAVIOUR Like all anglerfish, this fish has a lure. This is a short, thick spine with a fleshy end that is wiggled to look like a small fish. This tricks prey animals into coming too close to the batfish.

WALKING FISH
The batfish is not a great swimmer. Its pectoral fins are large and placed to the side. The fish uses them to 'walk' across the seabed.

01027 621021162 .83.9 0102 6210273301063

SIZE
25 CM
10 IN

RED LIPS

The bright red lips are easy to
spot in the gloom and they help
the male and female fish to
identify each other. The mouth
is wide to swallow prey.

CORAL REEFS

Exploring the coral reef is an unforgettable experience, as you move in the clear, sunlit waters in which the corals grow.

Coral reefs are the rainforests of the seas, brimming with fish and other animals, all dependent on each other in one way or another – perhaps for food, perhaps for protection, or even for a quick clean. There are brightly coloured clown fish darting in and out of sea anemones, lionfish armed with wicked-looking spines, thick-lipped triggerfish and coral-crunching parrotfish. Exotic sea slugs, known as nudibranchs, glide over the corals, while dainty crabs creep out of their hiding places. Large animals visit too, including manta rays looking for a cleaning station, and sharks hunting for their next meal of squid or cuttlefish.

It is difficult to believe that the reef was built by tiny animals, called coral polyps. Each coral polyp leaves behind a hard skeleton when it dies. This forms part of the reef, and over hundreds of years the reef gets slowly larger.

CORAL LIFE

More types of fish and other animals live on coral reefs than in any other marine habitat. The corals come in different shapes and colours. There are domed-, table-, fan- and antler-shaped corals, in colours ranging from blue and purple to red and orange.

BLACKFIN BARRACUDA

LATIN NAME *Sphyraena qenie*

The blackfin barracuda is a fierce predator on the coral reef. Its long, streamlined shape allows it to move quickly through the water. During the day, barracudas stay together in shoals, but at night they go their separate ways to hunt. Barracudas are inquisitive fish that come up to divers, but they may attack, especially in murky waters.

KILLER JAWS

The strong jaw of the barracuda allows it to grip its prey with great force. The fish's large mouth contains two sets of razor-sharp teeth to slice through flesh. Small prey is swallowed whole, while larger prey is cut to pieces.

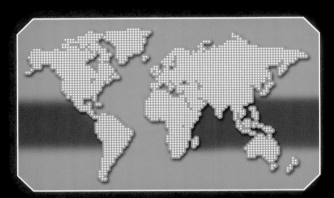

0 m
0 FT

100 m
330 FT

500 m
1640 FT

1000 m
3280 FT

2000 m
6560 FT

3000 m
9840 FT

4000 m
13120 FT

8000 m
26240 FT

12000 m
39360 FT

DEPTH GAUGE
5–50 M
(16–165 FT)

DISTRIBUTION The Red Sea and across the Indian and Pacific Oceans.

HABITAT The blackfin barracuda is found on coral reefs, where it prefers to hunt on the outer slopes of the reef.

DIET The barracuda hunts smaller fish, such as snapper and bream.

DESCRIPTION This fish has a long, silver body with 20 dark chevron bars. There are two dorsal fins, which lie far apart. The snout is long and the lower jaw sticks out.

RELATIVES There are 26 known species of barracuda. Some species can grow very large, such as the great barracuda, which can reach up to 2 m (6.6 ft) in length and 50 kg (110 lb) in weight.

BEHAVIOUR Most of the time, the barracuda hunts by ambushing its prey. It lies motionless in the water and then surprises its victim by lunging forwards. It may give chase over a short distance. Sometimes, barracudas hunt together, chasing smaller fish into the shallows, where they are easier to catch.

SIZE
1.5–1.7 M
5–5.6 FT

01027346210211 2754621027301063

SLEEK BODY

The barracuda has a streamlined body with two dorsal fins. As well as being ideal for speeding along, the barracuda's shape is well suited to surviving in the turbulent waters on the ocean side of a coral reef.

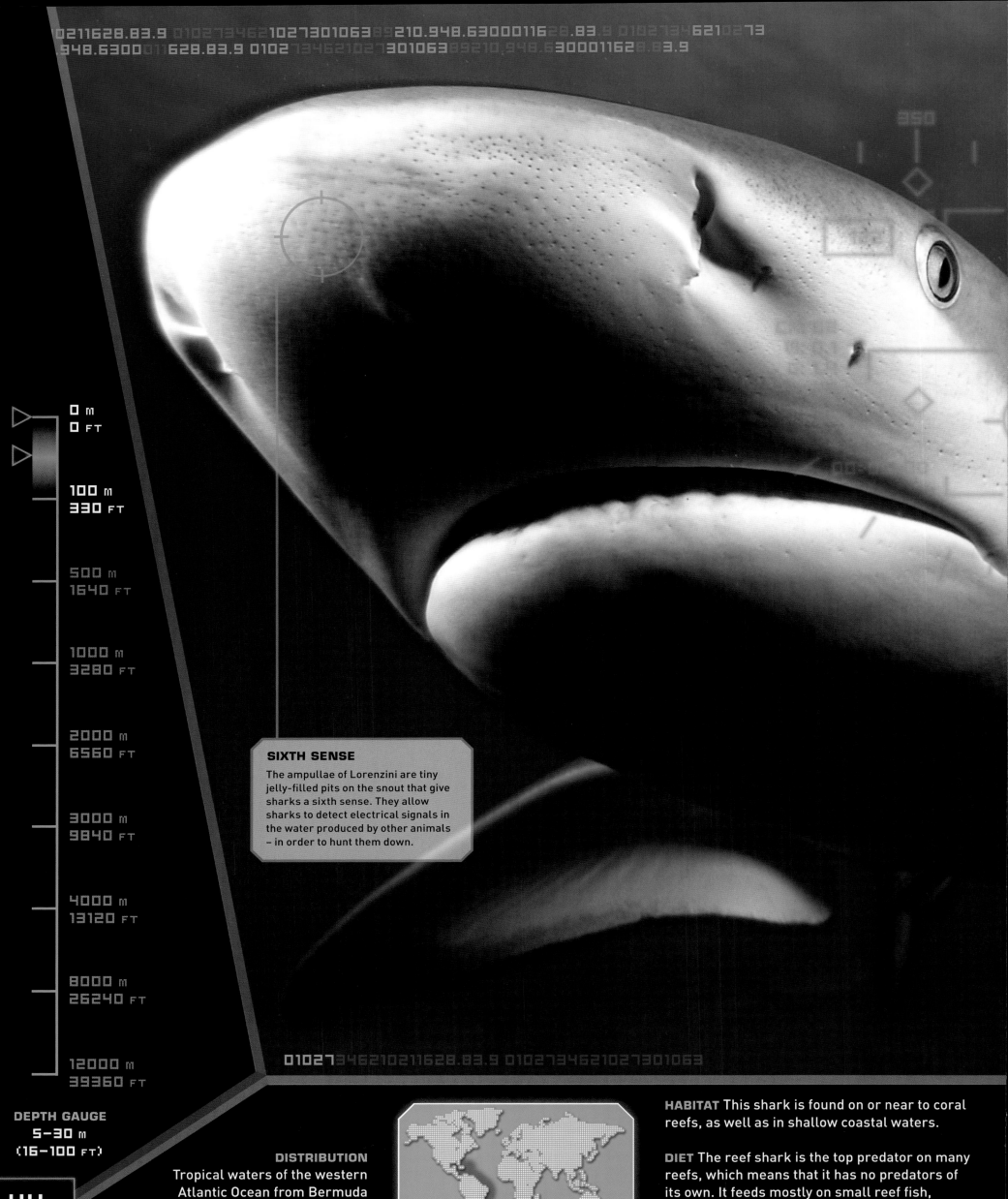

350

SIXTH SENSE

The ampullae of Lorenzini are tiny jelly-filled pits on the snout that give sharks a sixth sense. They allow sharks to detect electrical signals in the water produced by other animals – in order to hunt them down.

0 m	0 FT
100 m	330 FT
500 m	1640 FT
1000 m	3280 FT
2000 m	6560 FT
3000 m	9840 FT
4000 m	13120 FT
8000 m	26240 FT
12000 m	39360 FT

DEPTH GAUGE
5–30 m
(16–100 FT)

010273462102112628.83.9 01027346210273010638 01027346210273010638

44

DISTRIBUTION
Tropical waters of the western Atlantic Ocean from Bermuda and Florida south to Brazil.

HABITAT This shark is found on or near to coral reefs, as well as in shallow coastal waters.

DIET The reef shark is the top predator on many reefs, which means that it has no predators of its own. It feeds mostly on small reef fish, squid and cuttlefish.

CARIBBEAN REEF SHARK

LATIN NAME *Carcharhinus perezi*

Reef sharks are among the more common types of shark, although relatively little is known about them. This may be due to the fact that these sharks are nocturnal. During the day, reef sharks may be spotted resting on the seabed or in caves. They hunt for food at night. These sharks have been heavily fished and are now endangered. Fishing of the reef shark is banned in the USA.

SIZE
2–3 M
6.5–10 FT

BEHAVIOUR
The reef shark is viviparous, which means that the female gives birth to live young. Most sharks lay eggs, but the female reef shark keeps the eggs in her body. She gives birth to between four and six pups, each of which is about 75 cm (30 in) long.

THREAT DISPLAY
These sharks do not normally attack divers, but if they feel threatened, a reef shark will perform a threat display before attacking. The shark does a number of quick twists and turns, arches its back and lowers its pectoral fins.

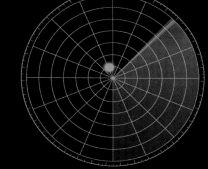

RANGEFINDER 1 m (3 FT)

DESCRIPTION The shark's upper surface is a grey to grey-brown colour, while the underside is white. There are two dorsal fins along this fish's back. The snout is short and broad, and the eyes are relatively large and round. The triangular teeth have a wide base and jagged edges.

RELATIVES There are 30 species of similar sharks belonging to the *Carcharhinus* genus. These species include the dusky, grey reef, sandbar, blacktip and silky sharks. Distinguishing features include the shape of the upper teeth, the length of the tip on the second dorsal fin, and the shape of the snout.

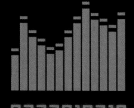

0273019210

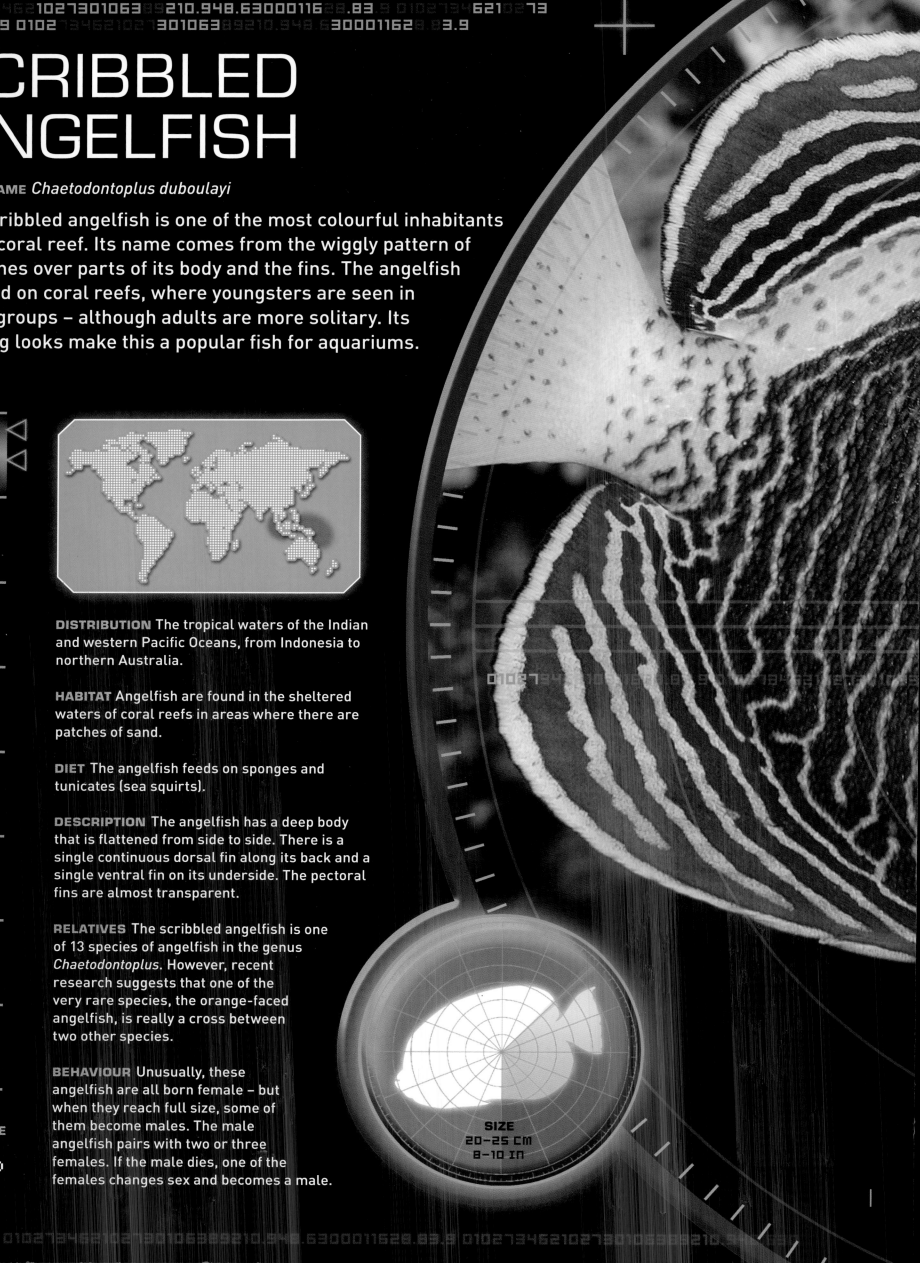

SCRIBBLED ANGELFISH

LATIN NAME *Chaetodontoplus duboulayi*

The scribbled angelfish is one of the most colourful inhabitants of the coral reef. Its name comes from the wiggly pattern of blue lines over parts of its body and the fins. The angelfish is found on coral reefs, where youngsters are seen in small groups – although adults are more solitary. Its striking looks make this a popular fish for aquariums.

0 m
0 FT

100 m
330 FT

500 m
1640 FT

1000 m
3280 FT

2000 m
6560 FT

3000 m
9840 FT

4000 m
13120 FT

8000 m
26240 FT

12000 m
39360 FT

DEPTH GAUGE
20–50 m
(65–165 FT)

DISTRIBUTION The tropical waters of the Indian and western Pacific Oceans, from Indonesia to northern Australia.

HABITAT Angelfish are found in the sheltered waters of coral reefs in areas where there are patches of sand.

DIET The angelfish feeds on sponges and tunicates (sea squirts).

DESCRIPTION The angelfish has a deep body that is flattened from side to side. There is a single continuous dorsal fin along its back and a single ventral fin on its underside. The pectoral fins are almost transparent.

RELATIVES The scribbled angelfish is one of 13 species of angelfish in the genus *Chaetodontoplus*. However, recent research suggests that one of the very rare species, the orange-faced angelfish, is really a cross between two other species.

BEHAVIOUR Unusually, these angelfish are all born female – but when they reach full size, some of them become males. The male angelfish pairs with two or three females. If the male dies, one of the females changes sex and becomes a male.

SIZE
20–25 CM
8–10 IN

SCRIBBLY LINES

The body of the angelfish has blue lines on black, with a speckled blue stripe through the head and a white and yellow bar behind. The dorsal and anal fins have brilliant blue stripes.

SPONGE-EATER

The scribbled angelfish has a small mouth with comb-like teeth that it uses to pull off bits of sponge. Angelfish are able to digest the tough flesh of sponges, animals that are not normally eaten by fish.

HARLEQUIN SHRIMP

LATIN NAME *Hymenocera picta*

This beautiful shrimp is a vicious inhabitant of the coral reef. It may be just a few centimetres long, but it attacks starfish many times its size. Working with its partner, the shrimp turns the starfish over and starts feeding. Sometimes the shrimps drag the starfish into a dark crevice, where they continue to feed on it for days, while it is still alive.

0 m
0 FT

100 m
330 FT

500 m
1640 FT

1000 m
3280 FT

2000 m
6560 FT

3000 m
9840 FT

4000 m
13120 FT

8000 m
26240 FT

12000 m
39360 FT

DEPTH GAUGE
5–50 m
(16–165 FT)

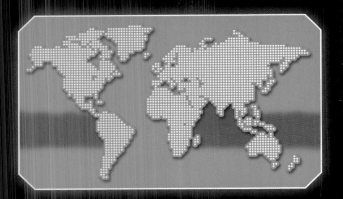

DISTRIBUTION From East Africa and the Red Sea to Indonesia, and from Australia across the Pacific to Hawaii and the Galápagos Islands.

HABITAT The harlequin is found on coral reefs where there are plenty of starfish.

DIET This specialized predator feeds only on starfish, such as the blue *Linckia*.

DESCRIPTION This colourful shrimp has a white to whitish-pink body, with blotches of pink, purple, blue or red, edged in blue. Its antennae are flattened and look a bit like leaves. Females have blue tips to their legs, but the males have transparent legs.

SUBSPECIES The appearance of this shrimp is quite variable. The harlequin shrimps of the Indian Ocean usually have blue and purple spots, while those living in the Pacific have pink, red or orange spots.

BEHAVIOUR The harlequin shrimp is always found in pairs. Each pair guards a territory on the reef. The female shrimp produces as many as 5000 eggs a year. Once she lays a batch of eggs, she looks after them, keeping them clean until they hatch.

WARNING COLOURS
The harlequin's bright colours are thought to be a warning to predators to stay away. If they do bite into the shrimp, they will discover that it tastes horrible after taking up poisons from the body of its starfish prey.

SIZE
3–5 CM
1.2–2 IN

48

PADDLE LEGS

The first pair of legs of these tiny shrimp are huge and look like paddles. They use the paddles to flip a starfish onto its back so that they can reach the tube feet – their favourite bit!

LIONFISH

LATIN NAME *Pterois volitans*

The lionfish is beautiful but deadly. This amazing creature, with its dramatic zebra-like stripes and large spiny fins, is among the most venomous of fish. It is the greediest of predators, eating about eight times its own weight in fish each year. The lionfish emerges to hunt at sunset, a time when daytime-feeding fish are returning to their hiding places and the night-feeding fish are becoming active.

VENOMOUS SPINES
When threatened, the fish lowers its head and spreads its venomous dorsal spines so that they point towards the threat. Its predators include the grouper and other lionfish.

0 m
0 FT

100 m
330 FT

500 m
1640 FT

1000 m
3280 FT

2000 m
6560 FT

3000 m
9840 FT

4000 m
13120 FT

8000 m
26240 FT

12000 m
39360 FT

DISTRIBUTION The Red Sea, Indian Ocean, central and western Pacific Ocean, and spreading into new areas, such as Florida and West Africa.

HABITAT The lionfish is found in lagoons and seaward-facing reefs, where it hides in rocky crevices during the day.

DIET This fearsome predator hunts for small fish, shrimp and crabs.

DESCRIPTION This lionfish has red-brown bands across the body and 13 dorsal spines down the back. The large pectoral fins have 14 long, feather-like rays. There is a tentacle-like growth above each eye, and a bony ridge on the cheek.

RELATIVES There are several lionfish species that look very similar. They differ only in the number of their poisonous spines. For example, *Pterois volitans* has 13 spines rather than 12.

BEHAVIOUR Most adults are solitary, with the males defending their territory. If a male lionfish meets another, the larger and more aggressive male turns a darker shade of red and raises its spines to look large and threatening. The weaker fish lowers its spines and swims away.

DEPTH GAUGE
3-50 m
(10-165 FT)

01027

SIZE
30-38 CM
12-15 IN

STEALTHY PREDATOR

The stripes of the lionfish provide camouflage on the reef, allowing it to swim up to unsuspecting fish from below and swallow them whole. Sometimes, it lies motionless in the water, waiting for fish to swim past.

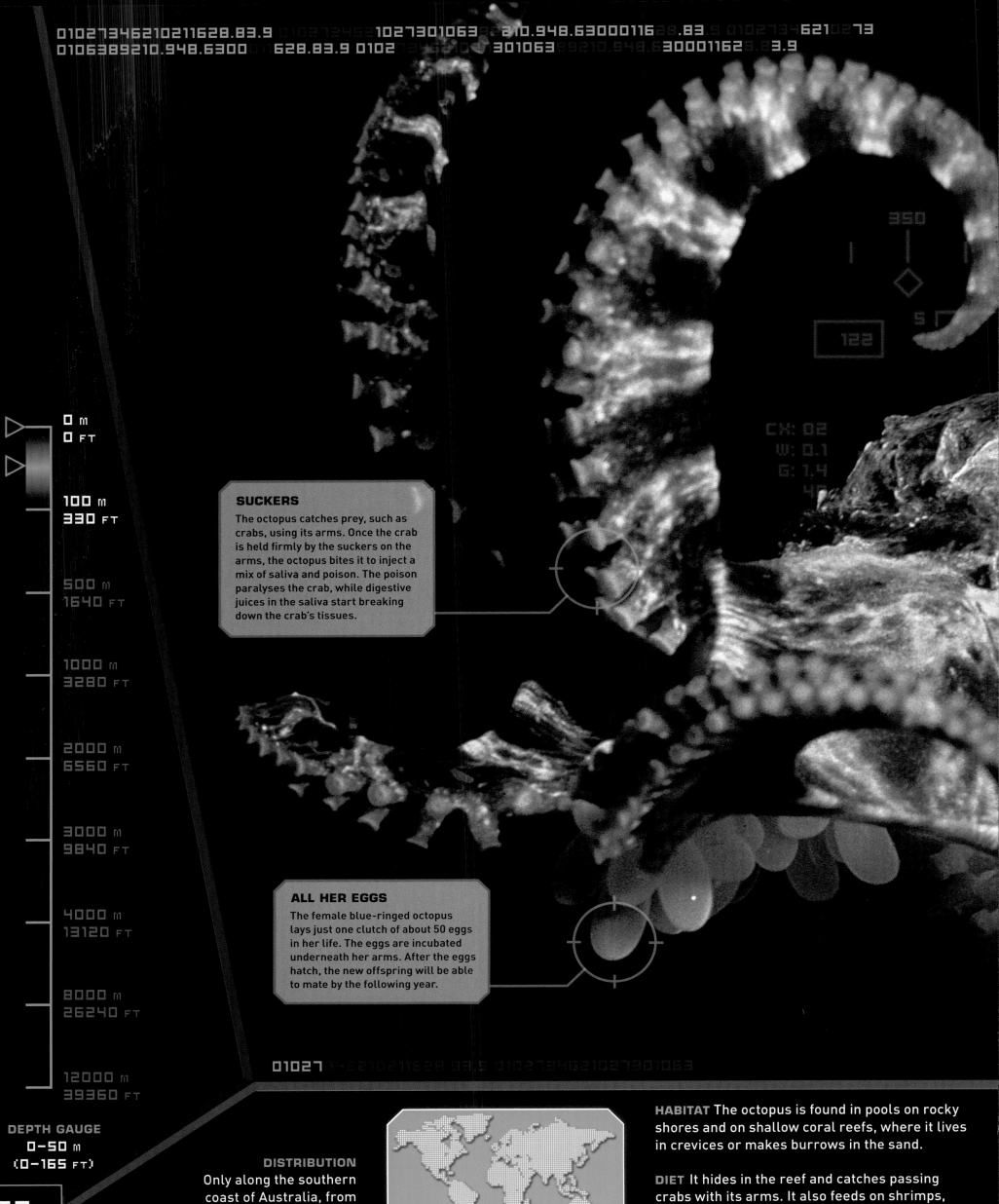

350

122

CX: 02
W: 0.1
G: 1.4

SUCKERS

The octopus catches prey, such as crabs, using its arms. Once the crab is held firmly by the suckers on the arms, the octopus bites it to inject a mix of saliva and poison. The poison paralyses the crab, while digestive juices in the saliva start breaking down the crab's tissues.

ALL HER EGGS

The female blue-ringed octopus lays just one clutch of about 50 eggs in her life. The eggs are incubated underneath her arms. After the eggs hatch, the new offspring will be able to mate by the following year.

0 m
0 FT

100 m
330 FT

500 m
1640 FT

1000 m
3280 FT

2000 m
6560 FT

3000 m
9840 FT

4000 m
13120 FT

8000 m
26240 FT

12000 m
39360 FT

01027

DEPTH GAUGE
0–50 m
(0–165 FT)

DISTRIBUTION
Only along the southern coast of Australia, from Western Australia to Victoria.

HABITAT The octopus is found in pools on rocky shores and on shallow coral reefs, where it lives in crevices or makes burrows in the sand.

DIET It hides in the reef and catches passing crabs with its arms. It also feeds on shrimps, bivalves and fish.

BLUE-RINGED OCTOPUS

LATIN NAME *Hapalochlaena maculosa*

The blue-ringed octopus is one of the deadliest animals in the ocean. At rest, the octopus is a pale yellow-brown colour, but when it is alarmed, iridescent blue rings appear over its body. The octopus has two types of venom. One is used to paralyse its prey, while the second, a more potent venom, is used for defence. Although it has enough venom to kill more than 20 people, this is not an aggressive octopus and it does not attack divers.

SQUIRTING SIPHON

A siphon is a tube just behind the head. To swim, the octopus squeezes its body to squirt a jet of water out of the siphon, and this propels the animal through the water. The females also use their siphon to direct a spray of water over their eggs to keep them free from parasites.

BEHAVIOUR

The male dies after mating. The female carries her eggs under her arms for about 50 days. During this time, she does not feed – and she dies after the eggs hatch. The hatchlings float in the water for about a month before dropping to the seabed.

SIZE
10 CM INCLUDING ARMS
4 IN

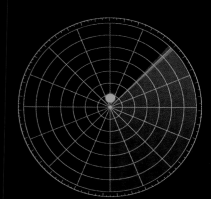

RANGEFINDER 10 cm (4 in)

DESCRIPTION This is a small octopus, with a body the size of a golf ball. It weighs less than 30 g (1 oz). Like all octopuses, it has eight arms with two rows of suction caps for gripping onto rocks, coral and prey. The surface of the body has a rough, wrinkled appearance.

RELATIVES There are 10 species of blue-ringed octopus, of which the southern blue-ringed octopus in the photograph is the largest and most common. The species differ slightly, for example in the size of their blue rings.

0273019210

ORANGE-STRIPED TRIGGERFISH

LATIN NAME *Balistapus undulatus*

The triggerfish gets its name from its long dorsal spine. The spine can be raised and locked into position by a smaller spine behind, making it difficult for the triggerfish to be swallowed by predators or pulled out of the crevices in which it rests. The spine is lowered when the second spine unlocks the first.

0 m
0 FT

100 m
330 FT

500 m
1640 FT

1000 m
3280 FT

2000 m
6560 FT

3000 m
9840 FT

4000 m
13120 FT

8000 m
26240 FT

12000 m
39360 FT

DEPTH GAUGE
5–50 m
(16–165 FT)

DISTRIBUTION Widely distributed in the tropical and subtropical waters of the Indian and Pacific Oceans, and the Red Sea.

HABITAT This fish likes deep coral lagoons and the seaward side of coral reefs.

DIET These omnivorous fish eat a range of foods, from algae and coral to sea urchins, crabs and fish.

DESCRIPTION The triggerfish has yellow or orange diagonal lines across its dark green to brown body. Blue and orange stripes run from the mouth to the pectoral fin. The tail fin and the base of the pectoral fins are orange.

RELATIVES The members of the triggerfish family are all brightly coloured fish that are found in warm and tropical waters. They include the titan triggerfish, which grows to 75 cm (30 in) and is very aggressive.

BEHAVIOUR The female triggerfish clears all the small rocks from an area of seabed to form a circular nest in the sand in which she lays a cluster of eggs. She fans the eggs with her pectoral fins to make sure they have plenty of oxygen.

SIZE
30 CM
12 IN

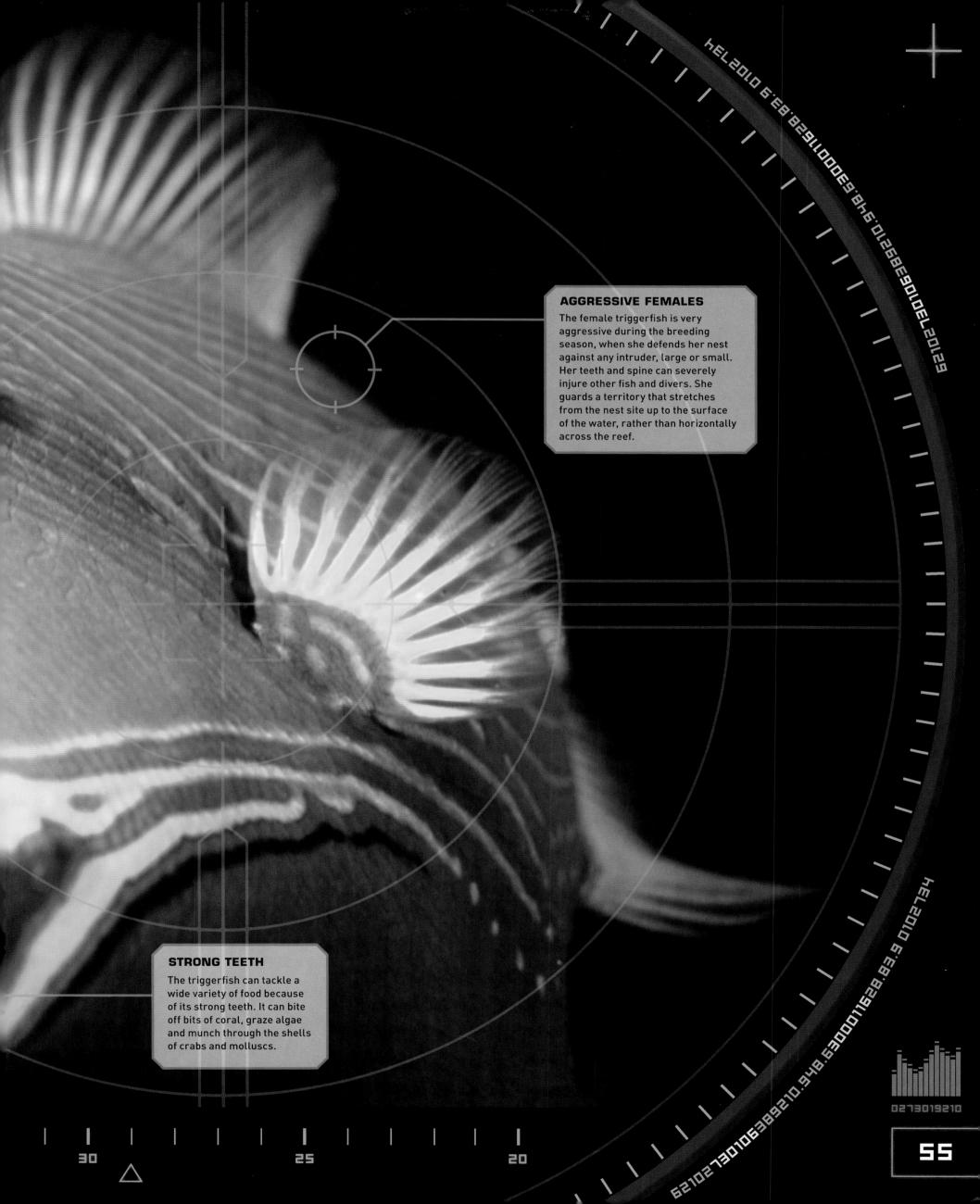

AGGRESSIVE FEMALES

The female triggerfish is very aggressive during the breeding season, when she defends her nest against any intruder, large or small. Her teeth and spine can severely injure other fish and divers. She guards a territory that stretches from the nest site up to the surface of the water, rather than horizontally across the reef.

STRONG TEETH

The triggerfish can tackle a wide variety of food because of its strong teeth. It can bite off bits of coral, graze algae and munch through the shells of crabs and molluscs.

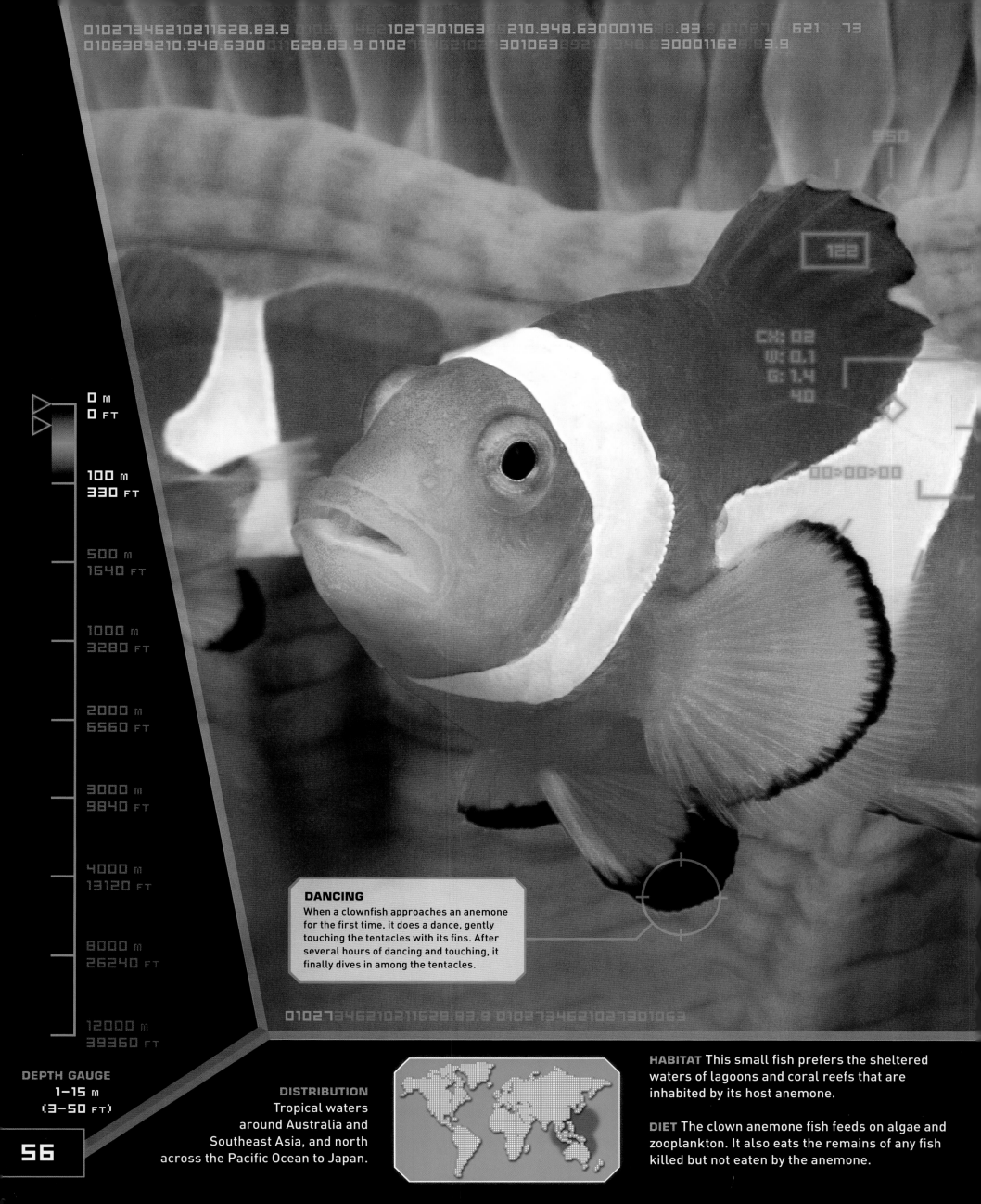

250

122

CX: 02
W: 0.1
G: 1.4
40

00>00>00

0 m
0 FT

100 m
330 FT

500 m
1640 FT

1000 m
3280 FT

2000 m
6560 FT

3000 m
9840 FT

4000 m
13120 FT

8000 m
26240 FT

12000 m
39360 FT

DANCING
When a clownfish approaches an anemone for the first time, it does a dance, gently touching the tentacles with its fins. After several hours of dancing and touching, it finally dives in among the tentacles.

0102734621021162B.83.9 01027346210273D1063

DEPTH GAUGE
1–15 m
(3–50 FT)

DISTRIBUTION
Tropical waters around Australia and Southeast Asia, and north across the Pacific Ocean to Japan.

HABITAT This small fish prefers the sheltered waters of lagoons and coral reefs that are inhabited by its host anemone.

DIET The clown anemone fish feeds on algae and zooplankton. It also eats the remains of any fish killed but not eaten by the anemone.

CLOWN ANEMONE FISH

LATIN NAME *Amphiprion ocellaris*

Few animals venture close to the stinging tentacles of the sea anemone. The clown anemone fish is different as it actually lives among the tentacles. Both the fish and the anemone benefit from this close relationship. The anemone provides the anemone fish with protection from its predators, while the fish keeps the anemone's tentacles free of parasites and chases away butterflyfish and turtles that prey on anemones.

IMMUNITY

Scientists are not sure how the fish is protected from the anemone's stings. There may be a chemical in the fish's mucus covering that gives protection. However, anemone fish lose their protection if they leave the anemone, so it may be a built-up immunity.

BEHAVIOUR

One anemone is host to a family of clown fish, which includes a breeding male and female pair and some young males. All clown fish are born male. When the female in a family dies, a male changes sex. The female lays eggs, then the breeding male guards them until they hatch.

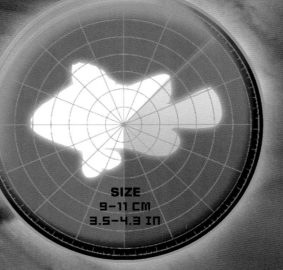

SIZE
9–11 CM
3.5–4.3 IN

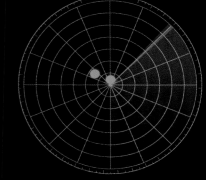

RANGEFINDER 5 CM (2 IN)

0273019210

DESCRIPTION This distinctive fish has an orange body with three broad white bars. The middle bar has a bulge that extends towards the head. The fins are edged in black. There are several very similar species that differ in the number of dorsal spines.

RELATIVES There are just under 30 species of clown fish belonging to the genus *Amphiprion*. They are found in the Indian and Pacific Oceans, where each species lives with a specific host anemone, which include *Heteractis* and *Stichodactyla* species.

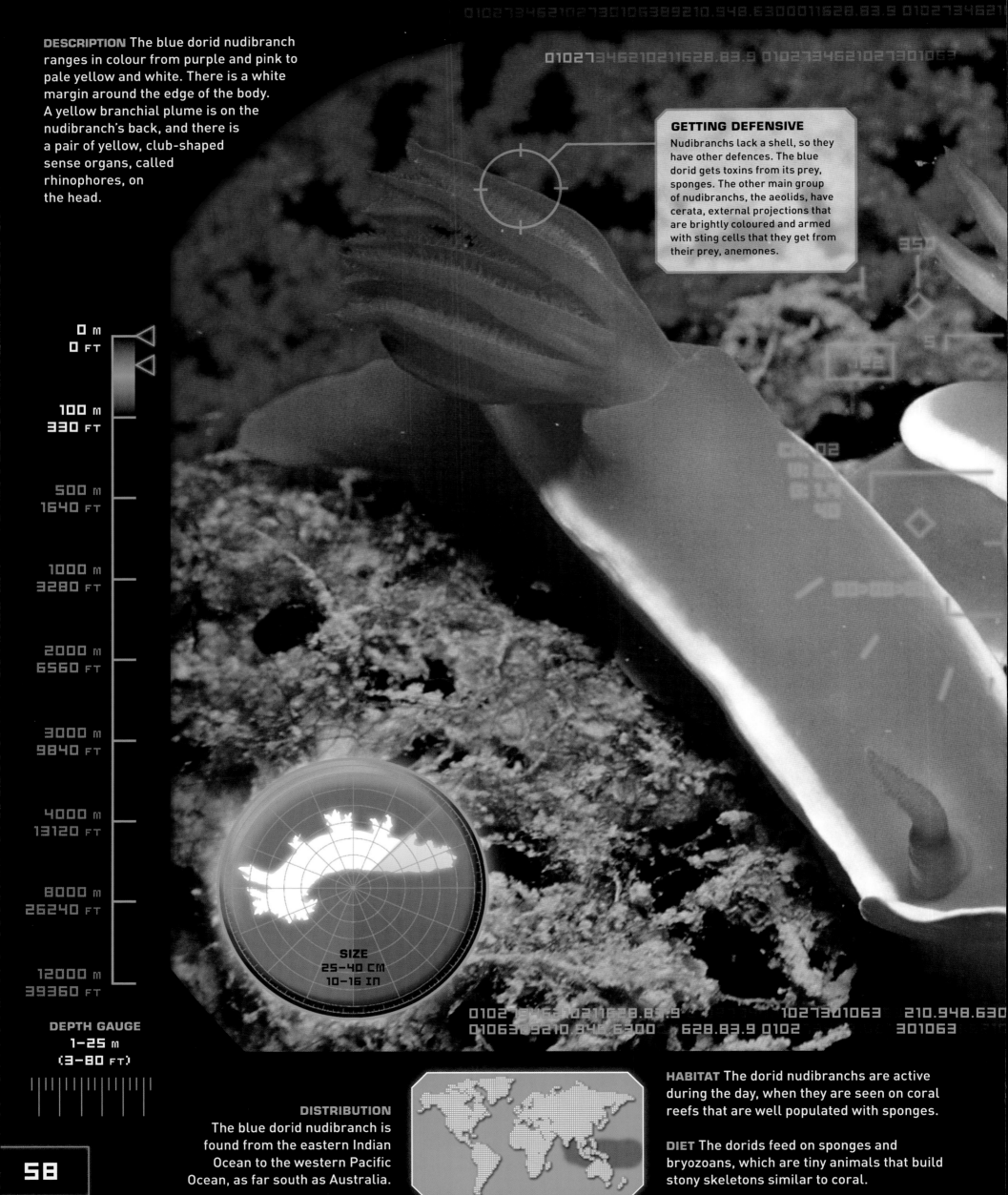

DESCRIPTION The blue dorid nudibranch ranges in colour from purple and pink to pale yellow and white. There is a white margin around the edge of the body. A yellow branchial plume is on the nudibranch's back, and there is a pair of yellow, club-shaped sense organs, called rhinophores, on the head.

GETTING DEFENSIVE
Nudibranchs lack a shell, so they have other defences. The blue dorid gets toxins from its prey, sponges. The other main group of nudibranchs, the aeolids, have cerata, external projections that are brightly coloured and armed with sting cells that they get from their prey, anemones.

0 m
0 FT

100 m
330 FT

500 m
1640 FT

1000 m
3280 FT

2000 m
6560 FT

3000 m
9840 FT

4000 m
13120 FT

8000 m
26240 FT

12000 m
39360 FT

DEPTH GAUGE
1–25 m
(3–80 FT)

SIZE
25–40 CM
10–16 IN

DISTRIBUTION
The blue dorid nudibranch is found from the eastern Indian Ocean to the western Pacific Ocean, as far south as Australia.

HABITAT The dorid nudibranchs are active during the day, when they are seen on coral reefs that are well populated with sponges.

DIET The dorids feed on sponges and bryozoans, which are tiny animals that build stony skeletons similar to coral.

NUDIBRANCH

LATIN NAME *Hypselodoris bullocki*

Nudibranchs, or sea slugs, are often called the jewels of the sea due to their bright colours. These colours warn other reef inhabitants that the nudibranch has a battery of defence systems in place, including poisons and sting cells. The pictured nudibranch, the blue dorid, belongs to the superfamily Doridoidea. These large nudibranchs have a branchial plume on their back that acts like a gill, extracting oxygen from the water.

LOOKING LIKE A MOLLUSC

Like all molluscs, nudibranchs have a muscular foot on which they glide over surfaces. Other mollusc-like features include a rasping tongue called a radula. Molluscs breathe with gills, which in the case of the blue dorid are external plumes.

BEHAVIOUR Nudibranchs are hermaphrodite, with both male and female sex organs. However, they do not fertilize their own eggs, but exchange sperm with other individuals. Then they lay an egg mass containing numerous eggs. The shape of the egg mass is dependent on the species.

SPECIES There are more than 3000 species of nudibranch, which are divided into two main groups: the dorids, such as the blue dorid, with a branchial plume, and the aeolids. The aeolids have long, narrow bodies, projections called cerata, and internal gills rather than branchial plumes.

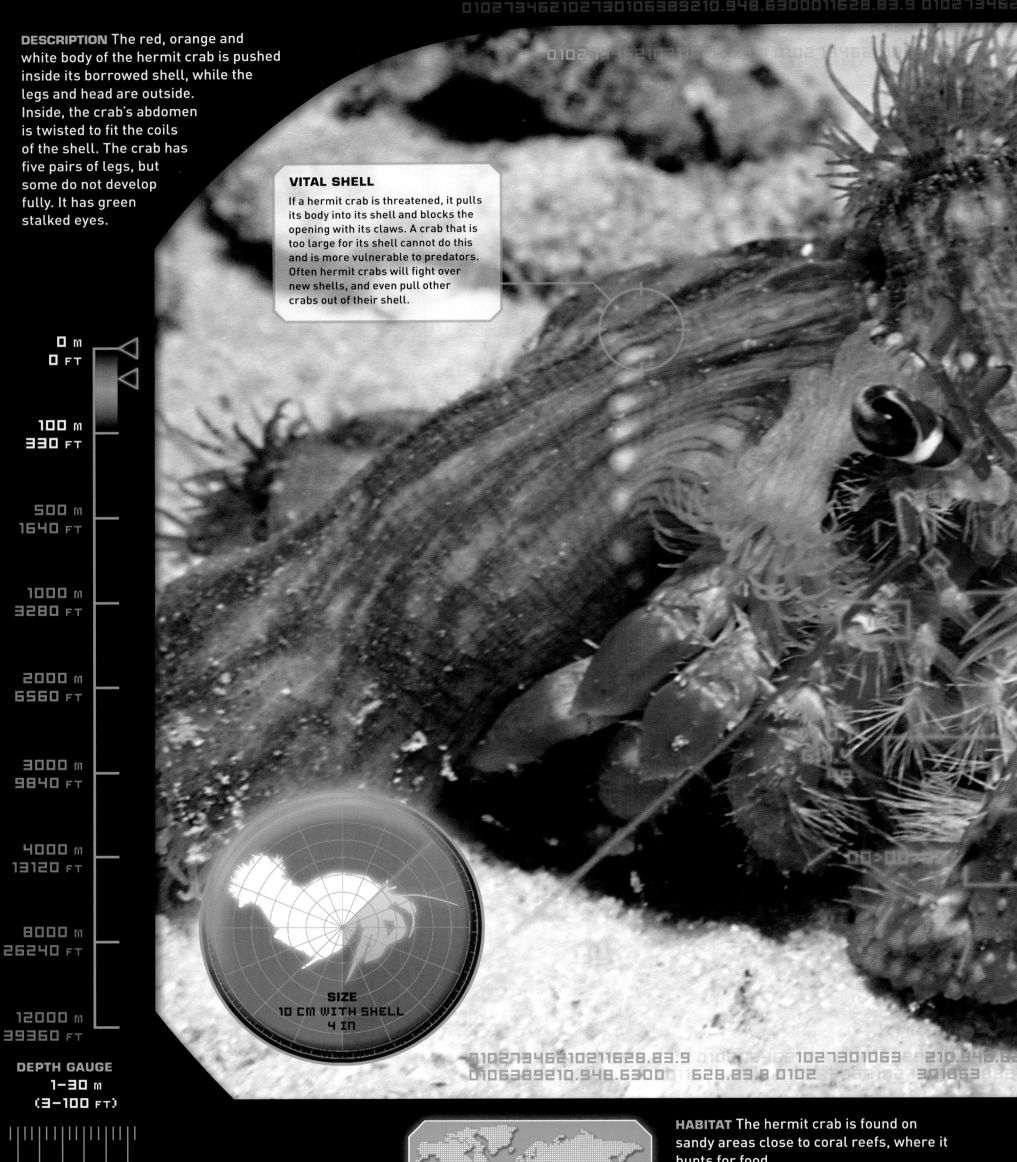

DESCRIPTION The red, orange and white body of the hermit crab is pushed inside its borrowed shell, while the legs and head are outside. Inside, the crab's abdomen is twisted to fit the coils of the shell. The crab has five pairs of legs, but some do not develop fully. It has green stalked eyes.

VITAL SHELL

If a hermit crab is threatened, it pulls its body into its shell and blocks the opening with its claws. A crab that is too large for its shell cannot do this and is more vulnerable to predators. Often hermit crabs will fight over new shells, and even pull other crabs out of their shell.

0 m
0 FT

100 m
330 FT

500 m
1640 FT

1000 m
3280 FT

2000 m
6560 FT

3000 m
9840 FT

4000 m
13120 FT

8000 m
26240 FT

12000 m
39360 FT

DEPTH GAUGE
1–30 m
(3–100 FT)

SIZE
10 CM WITH SHELL
4 IN

DISTRIBUTION
Across the Indian Ocean and the Pacific Ocean as far as Hawaii and Japan.

HABITAT The hermit crab is found on sandy areas close to coral reefs, where it hunts for food.

DIET The scavenging hermit crab eats almost anything it finds in its path, including fish, snails, worms and coral.

ANEMONE HERMIT CRAB

LATIN NAME *Dardanus pedunculatus*

Hermit crabs are crustaceans that carry around an abandoned shell. Unlike other crabs, the hermit has a soft abdomen that needs the protection of a second-hand snail or mollusc shell. The anemone hermit crab places several anemones on its shell. This is a mutualistic relationship, which means that both the crab and anemone benefit. The anemone camouflages and protects the crab, while the crab ensures that the anemone gets plenty of food.

ANEMONE PARTNERS

When crabs move from an old shell to a new one, they either move their anemones across or look for larger ones. Although the partnership is beneficial for both parties, a starving crab has been known to remove an anemone and eat it.

BEHAVIOUR The hermit crab relies on its borrowed shell for protection. The front edge of the crab's abdomen has a ridge that secures it inside the shell. The crab has to swap its shell for a larger one in order to grow bigger. Since whole, empty shells are sometimes in short supply, crabs often compete for them.

SPECIES There are about 800 species of hermit crab around the world, of which 44 belong to the genus *Dardanus*. The *Dardanus* hermit crabs all have associations with anemones. Although they are called hermit crabs, these creatures are more closely related to lobsters than to true crabs.

RANGEFINDER 5 cm (2 in)

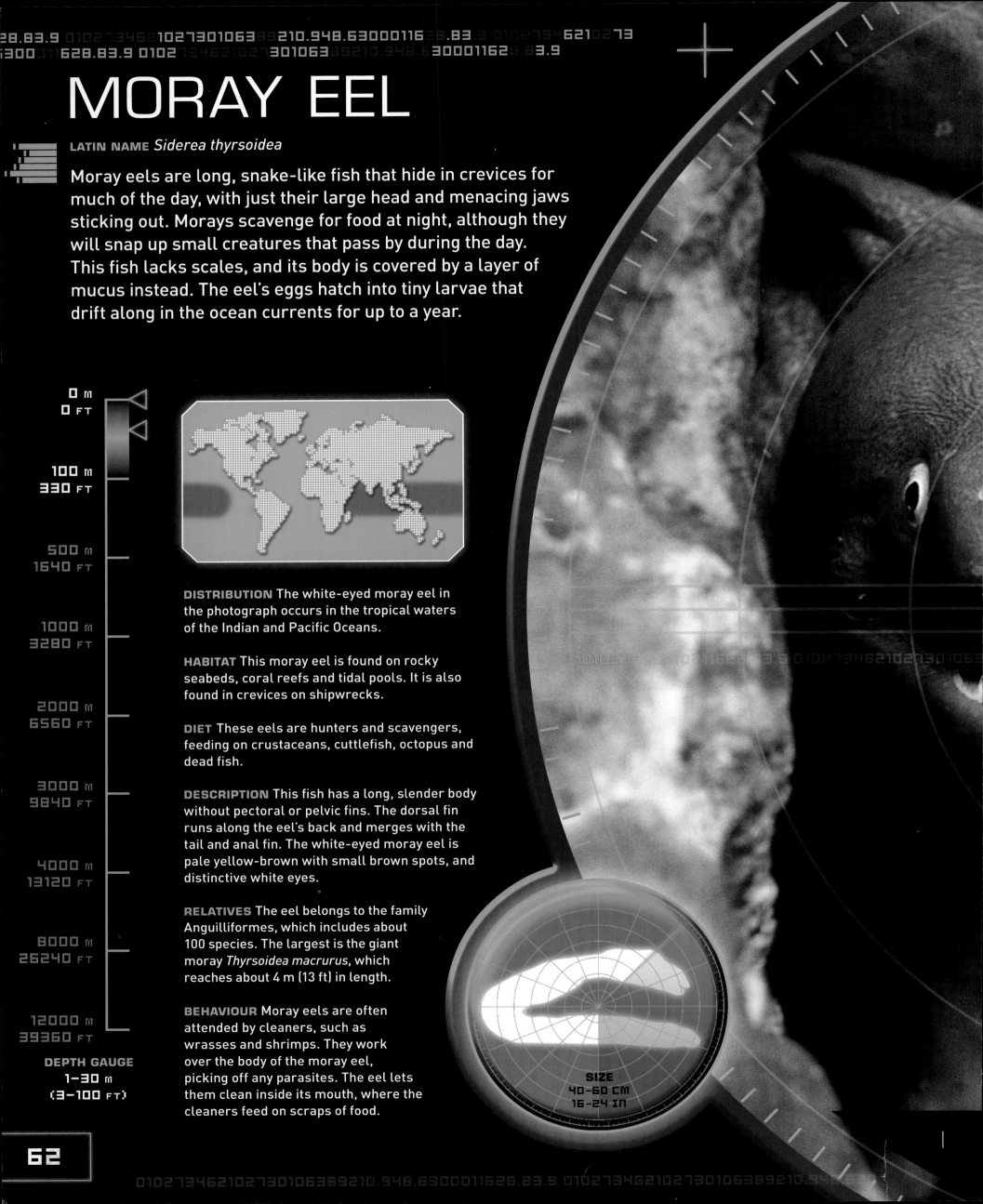

MORAY EEL

LATIN NAME *Siderea thyrsoidea*

Moray eels are long, snake-like fish that hide in crevices for much of the day, with just their large head and menacing jaws sticking out. Morays scavenge for food at night, although they will snap up small creatures that pass by during the day. This fish lacks scales, and its body is covered by a layer of mucus instead. The eel's eggs hatch into tiny larvae that drift along in the ocean currents for up to a year.

0 m
0 FT

100 m
330 FT

500 m
1640 FT

1000 m
3280 FT

2000 m
6560 FT

3000 m
9840 FT

4000 m
13120 FT

8000 m
26240 FT

12000 m
39360 FT

DEPTH GAUGE
1–30 m
(3–100 FT)

DISTRIBUTION The white-eyed moray eel in the photograph occurs in the tropical waters of the Indian and Pacific Oceans.

HABITAT This moray eel is found on rocky seabeds, coral reefs and tidal pools. It is also found in crevices on shipwrecks.

DIET These eels are hunters and scavengers, feeding on crustaceans, cuttlefish, octopus and dead fish.

DESCRIPTION This fish has a long, slender body without pectoral or pelvic fins. The dorsal fin runs along the eel's back and merges with the tail and anal fin. The white-eyed moray eel is pale yellow-brown with small brown spots, and distinctive white eyes.

RELATIVES The eel belongs to the family Anguilliformes, which includes about 100 species. The largest is the giant moray *Thyrsoidea macrurus*, which reaches about 4 m (13 ft) in length.

BEHAVIOUR Moray eels are often attended by cleaners, such as wrasses and shrimps. They work over the body of the moray eel, picking off any parasites. The eel lets them clean inside its mouth, where the cleaners feed on scraps of food.

SIZE
40–60 CM
16–24 IN

OPEN-MOUTHED

The moray eel's mouth is open most of the
time, so the mouth lining is camouflaged.
There are three rows of long, sharp teeth that
grasp large prey. There is also a second set of
jaws that stick out from the throat to grip prey
and drag it into the throat.

LYING IN WAIT

During the day, this eel hides in its hole and
waits for prey to come close. Then it lunges
forwards and grabs the animal. At night it
emerges from its hole to scavenge, relying
on its excellent sense of smell to find rotting
bodies on the seabed.

01027346210211628.83.9 0102734621027301063 9210.948.63000116 28.83 9 0102 746 621 0 73
0106389210.948.6300 1628.83.9 0102 7462102 7301063 9210.948 630001162 3.9

SPINY FINS

This grouper has a long dorsal fin with 11 dorsal spines, of which the third or fourth is the longest. There are also spines on the anal fin. The tail fin is large and rounded in shape, which helps the fish to lunge forward.

350

122

0 m
0 FT

100 m
330 FT

500 m
1640 FT

1000 m
3280 FT

2000 m
6560 FT

3000 m
9840 FT

4000 m
13120 FT

8000 m
26240 FT

12000 m
39360 FT

SIZE
1–2 m
3.3–6.6 FT

DEPTH GAUGE
5–30 m
(15–100 FT)

64

01027

DISTRIBUTION
The western Indian Ocean to the western Pacific Ocean, including Australia's Great Barrier Reef.

HABITAT It is usually found in lagoons and on the seaward side of coral reefs, where the seabed drops steeply into deep water.

DIET This predatory fish feeds on other fish, as well as cephalopods such as squid and octopus, and crustaceans including the spiny lobster.

POTATO GROUPER

LATIN NAME *Epinephelus tukula*

This fearless giant of a fish is named after the potato-shaped spots on its body. It is also called the potato cod and potato bass. This is an aggressive fish that guards its territory and chases off any intruders. It is a long-lived fish, but it grows very slowly, so it takes many years for numbers to recover after they have been overfished. The potato grouper is currently a protected species in Australian waters.

WIDE MOUTH

Potato groupers have a large, wide mouth with thick lips. The fish hides behind lumps of coral and lunges out at any passing prey. The prey is sucked into the grouper's mouth, with the help of powerful gill muscles, and is then swallowed whole.

BEHAVIOUR

During the breeding season, the adults gather in coral passes. These are narrow channels in the reef where there is fast-moving water flowing into and out of a lagoon. The currents carry the fertilized eggs away from the reef, where they would be eaten by other fish.

RANGEFINDER 1 m (3 FT)

DESCRIPTION The potato grouper is a large fish with a sturdy body. It weighs up to 100 kg (220 lb). It is grey-brown, with large dark brown spots and blotches on its body, although some individuals are almost black. On the head there are distinctive lines of small spots that radiate out from the eyes.

RELATIVES The potato grouper belongs to the subfamily Epinephelinae, which is made up of about 159 species. These species include the groupers, rock hinds and sea bass. Most of these fish are caught commercially, then sold for high prices in local markets.

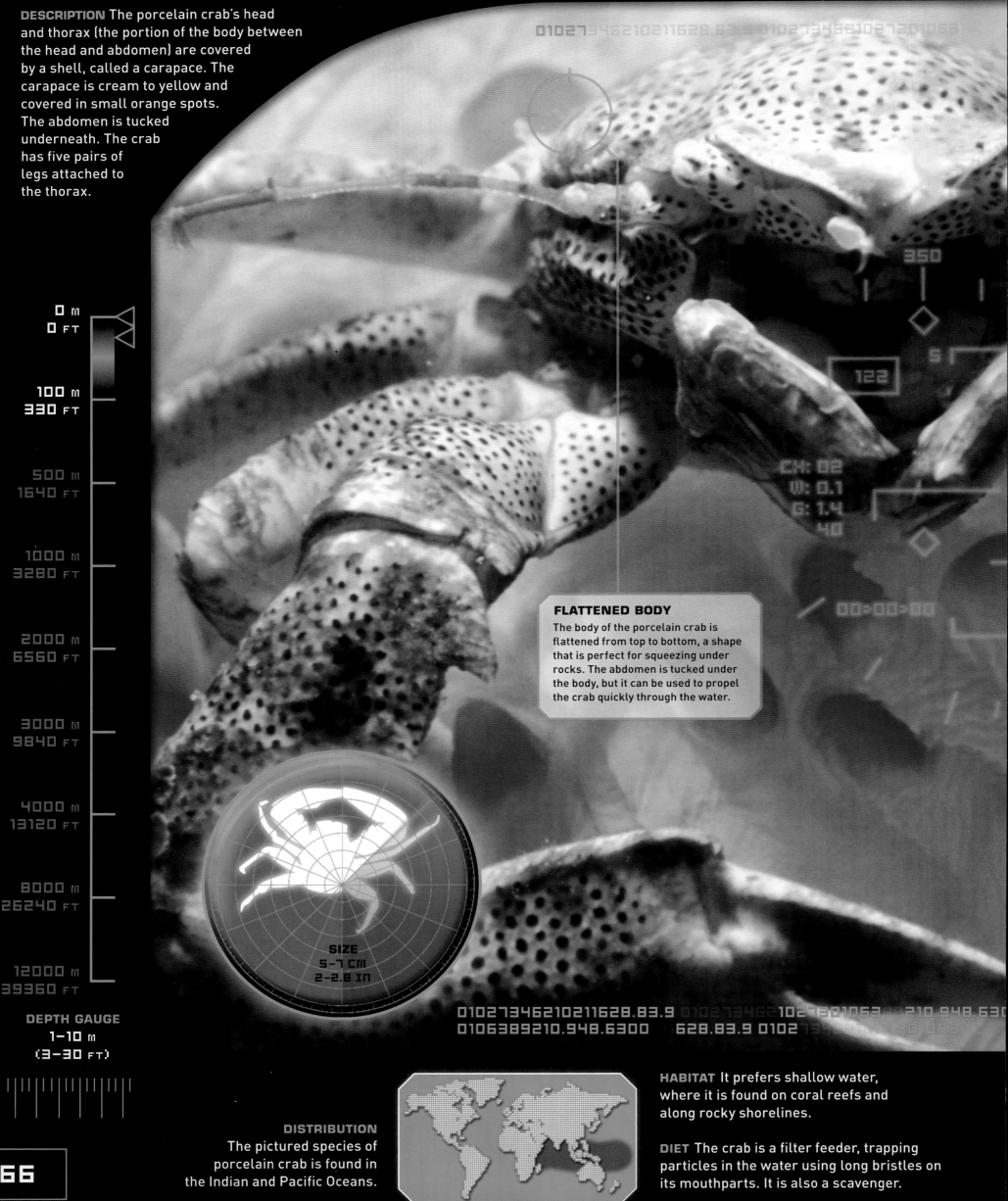

DESCRIPTION The porcelain crab's head and thorax (the portion of the body between the head and abdomen) are covered by a shell, called a carapace. The carapace is cream to yellow and covered in small orange spots. The abdomen is tucked underneath. The crab has five pairs of legs attached to the thorax.

FLATTENED BODY

The body of the porcelain crab is flattened from top to bottom, a shape that is perfect for squeezing under rocks. The abdomen is tucked under the body, but it can be used to propel the crab quickly through the water.

0 m
0 FT

100 m
330 FT

500 m
1640 FT

1000 m
3280 FT

2000 m
6560 FT

3000 m
9840 FT

4000 m
13120 FT

8000 m
26240 FT

12000 m
39360 FT

DEPTH GAUGE
1–10 m
(3–30 FT)

SIZE
5–7 CM
2–2.8 IN

DISTRIBUTION
The pictured species of porcelain crab is found in the Indian and Pacific Oceans.

HABITAT It prefers shallow water, where it is found on coral reefs and along rocky shorelines.

DIET The crab is a filter feeder, trapping particles in the water using long bristles on its mouthparts. It is also a scavenger.

PORCELAIN CRAB

LATIN NAME *Neopetrolisthes maculata*

The tiny porcelain crab gets its name from the way its legs break off easily – as if they were made of fine china – and then regrow. The porcelain crab is not a true crab, but an animal that has evolved to look like a crab. These animals are usually found in pairs, each pair living with a large anemone. The crabs clean the anemone, while in return, they are protected by the anemone's stinging tentacles.

TEN LEGS

Both true crabs and porcelain crabs are decapods, or ten-legged animals. In porcelain crabs, the first pair of legs end in claws that are used to pick up food and move objects, the next three pairs are used for walking, and the last pair is tucked behind the carapace. A true crab has four pairs of walking legs.

BEHAVIOUR The female porcelain crab carries around a batch of up to 1600 eggs in the brood flap on her abdomen until they hatch. The tiny larval crabs float in the ocean currents for several weeks before sinking to the seabed and developing into adults.

SPECIES Although it is called a crab, the porcelain crab belongs to a completely different family from the true crabs. It is in the Porcellanidae family, which includes the hermit crabs and squat lobsters. There are just over 30 species of porcelain crabs found around the world.

RANGEFINDER 2 cm (1 in)

GREENTHROAT PARROTFISH

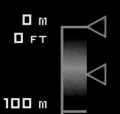

LATIN NAME *Scarus prasiognathos*

The parrotfish gets its name from its large, beak-like teeth. This fish is essential to the health of the coral reef, as its constant grazing stops algae from smothering the coral. At night, the parrotfish squeezes into a crevice and secretes, or releases, a cocoon of mucus around its body. This cocoon stops it being seen by predators.

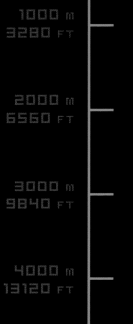

0 m
0 FT

100 m
330 FT

500 m
1640 FT

1000 m
3280 FT

2000 m
6560 FT

3000 m
9840 FT

4000 m
13120 FT

8000 m
26240 FT

12000 m
39360 FT

DEPTH GAUGE
3–50 m
(10–165 FT)

DISTRIBUTION The tropical waters of the Indian and Pacific Oceans.

HABITAT This parrotfish prefers the outer reefs, but it may also be seen in shallower, sheltered waters.

DIET The parrotfish grazes mostly on algae but will also bite off lumps of coral.

DESCRIPTION Parrotfish have deep bodies that are covered in large, thick scales. The males are brightly coloured in green and blue, with orange behind the head and on the fins. The female is much less colourful, with a green-brown body and orange-brown fins and head.

RELATIVES Parrotfish belong to the genus *Scarus*, which contains just over 80 species. Biologists believed there to be many more species until it was discovered that the females had been identified as separate species.

BEHAVIOUR The fish are born hermaphrodite, with both male and female sex organs, and they change sex as they develop. The male guards a small group of females, known as his harem. If he dies, one of the females changes sex to replace him.

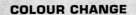

COLOUR CHANGE
Juvenile fish, both male and female, have a drab colour. Males gain their bright colours as they get older. Primary, or dominant, males are those that have never been female. However, most males were once female.

SIZE
50–70 CM
20–28 IN

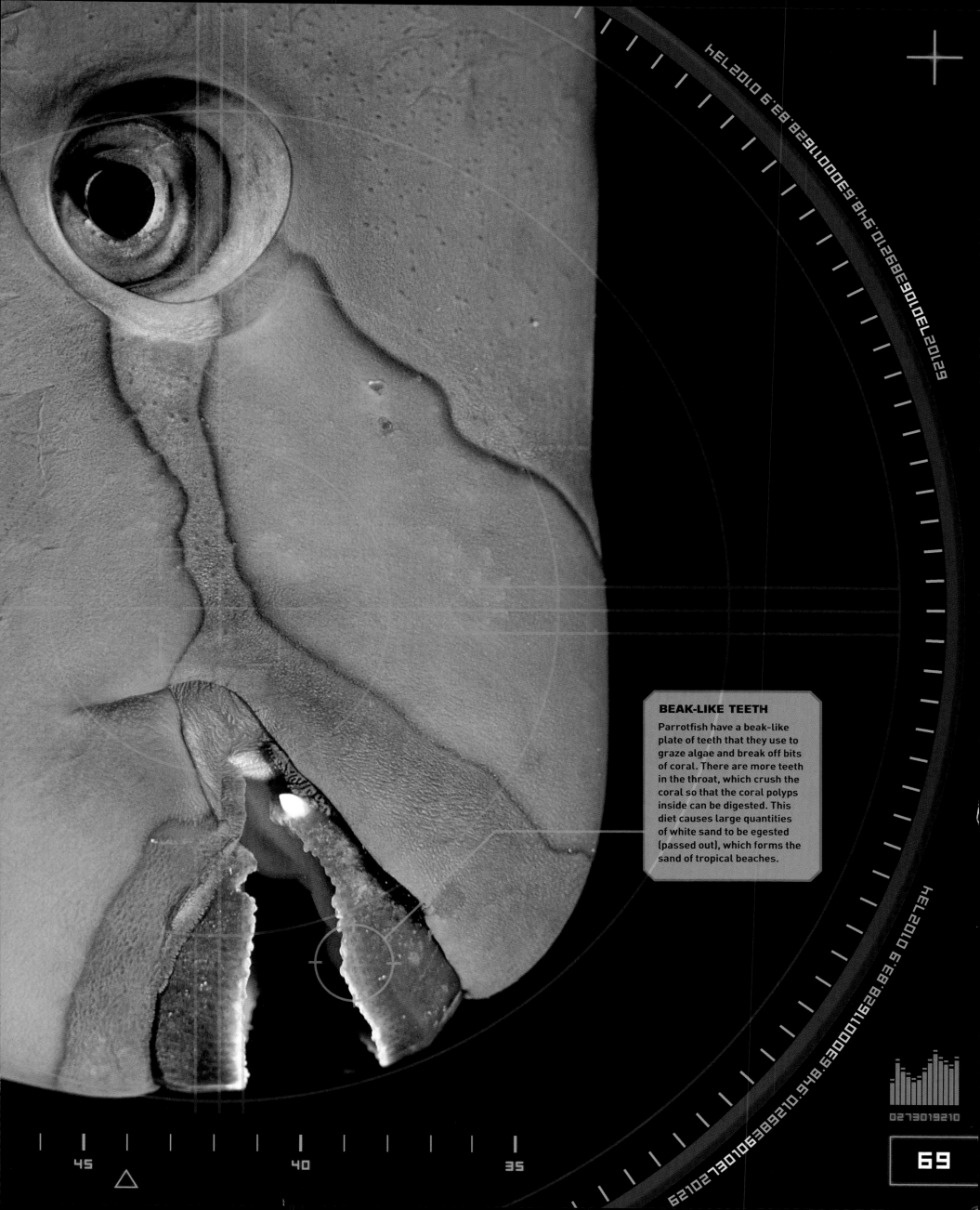

BEAK-LIKE TEETH

Parrotfish have a beak-like plate of teeth that they use to graze algae and break off bits of coral. There are more teeth in the throat, which crush the coral so that the coral polyps inside can be digested. This diet causes large quantities of white sand to be egested (passed out), which forms the sand of tropical beaches.

CEPHALIC LOBES

The odd-looking flaps, or lobes, beside the mouth funnel water rich in plankton into the mouth. This water is then passed back over the gills. As water moves through the gill bars, small particles of food are filtered out.

SIZE
6-7 M
20-23 FT

0 m
0 FT

100 m
330 FT

500 m
1640 FT

1000 m
3280 FT

2000 m
6560 FT

3000 m
9840 FT

4000 m
13120 FT

8000 m
26240 FT

12000 m
39360 FT

DEPTH GAUGE
5-50 m
(15-165 FT)

DISTRIBUTION
Tropical waters, between 35°N and 35°S, all around the world.

HABITAT Mantas prefer shallow water close to coral reefs. However, they are also seen in the open ocean.

DIET These giant fish feed on plant and animal plankton, including larval fish and small crustaceans.

MANTA RAY

LATIN NAME *Manta birostris*

The manta is the largest of the rays, a type of cartilaginous fish related to sharks. Cartilaginous fish have skeletons made of flexible cartilage rather than bone. Despite its menacing appearance, the manta ray feeds peacefully on tiny floating plankton. It glides through the ocean by flapping its wing-like fins up and down. Individuals can be identified by the pattern of blotches and scars on their underside.

PROTECTIVE LAYER

Most rays have a rough skin, but mantas have a much smoother skin that is covered by a thick layer of mucus. If they are touched by a human diver, the mucus is removed and they are likely to pick up infections.

BEHAVIOUR

Small fish called remoras are often attached to the underside and the head of a manta. These fish feed on parasites and eat food that spills out of the mouth. The manta also visits cleaning stations on reefs, where small fish and shrimp clean its skin of parasites.

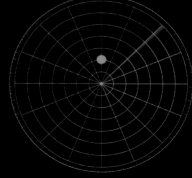

RANGEFINDER **5 m (16 FT)**

DESCRIPTION Manta rays are large, weighing up to 1300 kg (2900 lb). They are dark brown to black on top, and white underneath. There are two large flaps, called cephalic lobes, on either side of the mouth. Mantas lack the large teeth found in other species of ray.

SPECIES Recent scientific studies on manta rays have shown that there are two species, not just one as was first believed. One species is smaller and stays within one area, such as a coral reef, while the second species is larger and migrates over large distances each year.

0273019210

AMBIGUOUS SEA SPIDER

LATIN NAME *Pseudopallene ambigua*

The ambiguous sea spider is a small, long-legged marine animal, barely a centimetre or so across. It gets its name from its spider-like appearance but it is not a true spider. It feeds by inserting its proboscis – an extended mouthpart – into its prey, pouring over digestive juices, and then sucking the food up.

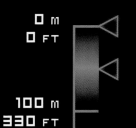

0 m
0 FT

100 m
330 FT

500 m
1640 FT

1000 m
3280 FT

2000 m
6560 FT

3000 m
9840 FT

4000 m
13120 FT

8000 m
26240 FT

12000 m
39360 FT

DEPTH GAUGE
1–50 m
(3–165 FT)

DISTRIBUTION The coastal waters of Australia.

HABITAT The ambiguous sea spider prefers shallow waters, where it is found on coral reefs and rocky seabeds. Some of the larger species are found in the deepest parts of the oceans.

DIET These carnivorous animals feed on anemones, sponges, marine worms and tiny coral-like bryozoans. They also scavenge on dead animals.

DESCRIPTION The ambiguous sea spider has a small yellow body with four pairs of long legs. The head bears four simple eyes; a pair of chelicerae, or pointed mouthparts, for gripping food; a pair of sensory palps; and a sucking proboscis.

RELATIVES There are about 1300 species of sea spider or Pycnogonids. There are many differences – such as the proboscis – between sea spiders and true spiders.

BEHAVIOUR The female lays her mass of eggs directly onto the egg-carrying leg of the male, who carries them until they hatch. This improves the eggs' chances of survival, but there is no parental care after they hatch.

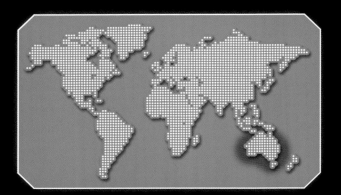

SIZE
1–20 MM
0.04–0.8 IN

BREATHING

With its long legs, the surface area of this tiny sea spider is large compared with its volume. This means that oxygen can diffuse – or be absorbed – through the external skeleton into the body. There is no need to have lungs or gills.

MOORISH IDOL

LATIN NAME *Zanclus cornutus*

This fish may get its name from the peoples of North Africa, whom medieval Europeans called Moors. These peoples were said to believe that the fish brings happiness. The idol is also called the common scythe, after its sickle-shaped dorsal fin. This fin extends behind the fish, but gets shorter with age. Moorish idols are active on the reef during the day, and rest at night, when their colours become more drab and less noticeable.

0 m
0 FT

100 m
330 FT

500 m
1640 FT

1000 m
3280 FT

2000 m
6560 FT

3000 m
9840 FT

4000 m
13120 FT

8000 m
26240 FT

12000 m
39360 FT

DEPTH GAUGE
3–180 m
(10–600 FT)

74

POKING SNOUT
The Moorish idol has a long, thin snout that ends in a small, round mouth. This is the ideal shape for poking into cracks and crevices on the reef in order to reach its favourite food, algae and sponges.

DISTRIBUTION Across the Indian and Pacific Oceans, from East Africa and the Red Sea to the west coast of Central and South America.

HABITAT The Moorish idol prefers lagoons, rocky cliffs and flat coral reefs, where there are crevices in which to hide.

DIET It feeds on algae and a range of invertebrates found on the seabed, such as sponges and sea squirts.

DESCRIPTION The deep body is flattened from side to side and is marked with bands of black, yellow and white. The pectoral fins are small, while the dorsal fin has a long trailing spine. The *cornutus* part of the idol's Latin name means 'horns' and refers to the lumps over the fish's eyes, which are larger in the male.

RELATIVES The Moorish idol may look like a butterflyfish or an angelfish, but it is more closely related to the plainer sturgeon.

BEHAVIOUR The larvae spend several months drifting in the currents, during which time they can travel long distances. When they are about 7 cm (2.8 in) long, they change into free-swimming fish.

SIZE
15–23 CM
6–9 IN

PAIRED FOR LIFE

The Moorish idol mates for life. Each pair keeps apart from other Moorish idols on the reef. The males become very aggressive during the breeding season. Young Moorish idols may be seen in small shoals on the reef.

POLAR REGIONS

 Our submersible edges through the sea ice of the Arctic Ocean before heading towards the South Pole and the chilly waters of the Southern Ocean.

Not only do animals living in polar waters have to cope with the tides, but they have to deal with the extreme climate too. During the summer months, the sun never sets at the Poles, but as autumn approaches, the days get shorter. During the polar winter, the sun never rises and it is dark all the time. It is incredibly cold – throw a bucket of water and it freezes almost before it touches the ground. The ocean is covered by a thick layer of ice.

During the summer months, many animals visit the polar waters to feed. The cold water supports vast amounts of plankton and krill, which attract the largest animals in the ocean, the whales. There are predators too, including the orca and the leopard seal. When winter approaches, the summer visitors migrate to escape the cold. Only the toughest remain: animals such as, in the Antarctic, the emperor penguin and, in the Arctic, the walrus.

02734621027301063892I0.948.63000II628.83.9 0I02734638921O.
8.6302I02730I063892I0.948.630

ICY OCEANS

The North Pole is covered by the Arctic Ocean. It is so cold there that the ocean is partly frozen over all year round – and almost entirely coated in ice in winter. The South Pole is covered by the white continent of Antarctica, which is surrounded by the Southern Ocean. In winter this vast ocean is almost totally covered in ice.

HUMPBACK WHALE

LATIN NAME *Megaptera novaeangliae*

The humpback whale is one of the largest of the baleen whales. These whales feed by filtering water containing tiny creatures through the baleen plates in their mouth. The humpback gets its name from the way its back arches as it dives. It comes to the surface to breathe about every 10 minutes, although it can stay underwater for up to 30 minutes. These are acrobatic mammals that can leap right out of the water.

0 m	**0** FT
100 m	**330** FT
500 m	**1640** FT
1000 m	**3280** FT
2000 m	**6560** FT
3000 m	**9840** FT
4000 m	**13120** FT
8000 m	**26240** FT
12000 m	**39360** FT

DEPTH GAUGE
5–210 m
(15–690 FT)

LONG FLIPPERS

The humpback has the longest flippers – in relation to its size – of any whale. These help the whale to steer and turn as it ploughs through the oceans. Scientists think that they also help with keeping the whale's temperature steady.

DISTRIBUTION
Found across all the oceans of the world at different times of year.

HABITAT These whales are found in the surface layer of the open ocean, and in shallow coastal waters, where they give birth.

DIET Humpbacks eat up to 2500 kg (5500 lb) of food a day, feeding on krill and small fish, especially herring, mackerel and capelin.

SIZE
12-16 M
40-50 FT

BEHAVIOUR
The haunting song
the male humpbac
made up of a patter
of squeaks, roars an
groans. It lasts for
about 20 minutes, bu
is repeated for a coup
of hours. All the whale
in one ocean sing the
same song, although
each year the song
changes slightly.

THROAT GROOVES
There are between 16 and 35 grooves,
running from the lower jaw to just
behind the pectoral fin. These allow
the throat to expand when the whale
swallows water. This water is forced
through the baleen plates in the mouth.

RANGEFINDER 10 m (33

DESCRIPTION The whale varies in colour
from grey and white to black and mottled.
The head is covered in lumps called
tubercles. There are long wing-like
flippers and a large tail fluke. Individuals
can be identified by the patterns on the
tail and the flippers.

SUBSPECIES The humpback is most
closely related to the grey and fin
whales, which are also baleen whales. There
are four main groups, or populations, within the
humpback species. Those whales living in the North
Pacific, Atlantic and Southern Oceans migrate each
year, but those living in the Indian Ocean do not.

DESCRIPTION The penguin has a sleek body with flipper-like wings. There are black feathers on its head, back and upper wings, and white feathers on the front of the body and undersides of the wings. A yellow-orange patch is on either side of the head. The tongue is covered in barbs to stop fish escaping from the mouth.

0 m
0 FT

100 m
330 FT

500 m
1640 FT

1000 m
3280 FT

2000 m
6560 FT

3000 m
9840 FT

4000 m
13120 FT

8000 m
26240 FT

12000 m
39360 FT

DEPTH GAUGE
0–550 m
(0–1800 FT)

EXPERT SWIMMER
The penguin's streamlined body slips easily through the water. Its powerful flippers propel the bird at speeds of up to 4–6 kph (2.5–4 mph). The emperor usually dives to depths of 150–250 m (500–800 ft), for up to 20 minutes, but it can descend to 500 m (1600 ft) or more.

SIZE
1–1.3 m
3.3–4.3 FT

DISTRIBUTION
Found only in the Antarctic, between latitudes 66° and 77° south.

Antarctic

HABITAT The adults go to sea for several months of the year, and come onto the pack ice along the coast to breed.

DIET This predatory penguin feeds mostly on fish, such as the Antarctic silverfish, as well as on squid, krill and other crustaceans.

EMPEROR PENGUIN

LATIN NAME *Aptenodytes forsteri*

The emperor is the largest of the penguins, standing up to 1.3 m (4.3 ft) high. A flightless bird, it is quite clumsy on land, but once in the sea it is a graceful swimmer. This penguin is one of the few animals that can survive winter in the Antarctic, where temperatures can plummet to –60°C (–76°F). Each year the adult penguins trek up to 100 km (60 miles) to reach their breeding grounds on the pack ice.

KEEPING WARM

In order to survive the extreme Antarctic cold, the emperor penguin is covered in a thick layer of feathers. Further insulation comes from a layer of fat beneath the skin, called blubber, which is up to 3 cm (1.2 in) thick.

BEHAVIOUR Breeding starts in winter, when the female lays one egg which is then incubated (kept warm) by the male. The females return to the sea to feed, while the males huddle together to keep warm. When the chick hatches, the female returns with food. Then the pair take it in turns to look after the chick.

RELATIVES There are 17 species of penguin, which are found only in the southern hemisphere. Five species occur in the Antarctic: the emperor, chinstrap, Adélie, gentoo and macaroni. The smallest penguin is the fairy or little blue penguin, which is just 40 cm (16 in) tall.

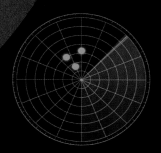

RANGEFINDER **6** m (20 FT)

WOLF FISH

LATIN NAME *Anarhichas lupus*

The fearsome wolf fish, or seawolf, gets its name from its fang-like teeth. This fish is found in the nearly freezing waters of the North Atlantic Ocean. It has a natural antifreeze circulating in its blood that stops its tissues from freezing in the cold water. This is a solitary fish that lives for about 20 years. The number of wolf fish is falling due to overfishing, especially in the coastal waters of North America.

LIKE AN EEL
The wolf fish has large, rounded pectoral fins on its sides. It has no pelvic fins, which would normally be found on the underside of a fish. The wolf fish swims slowly over the seabed, throwing its body from side to side in the manner of an eel.

0 m
0 FT

100 m
330 FT

500 m
1640 FT

1000 m
3280 FT

2000 m
6560 FT

3000 m
9840 FT

4000 m
13120 FT

8000 m
26240 FT

12000 m
39360 FT

DEPTH GAUGE
1–600 m
(3–1970 FT)

DISTRIBUTION The North Atlantic Ocean, especially along the coasts of Scandinavia, Scotland, Iceland and Greenland, and south along the North American coast to Cape Cod.

HABITAT It is found on rocky seabeds where there are rocks, crevices and small caves in which to hide, with just the head sticking out.

DIET The wolf fish eats a range of invertebrates living on the seabed, including molluscs such as whelks, clams and cockles, hermit crabs, green crabs, sea urchins and starfish.

DESCRIPTION This is an eel-like fish with a large head. It weighs up to 23 kg (50 lb). It is olive green to blue-grey, with dark bars on the body and dorsal fin. The spiny dorsal fin extends all the way along the back.

RELATIVES The wolf fish belongs to the family Anarhichadidae, in which all the species have the long, spiny dorsal fin and large teeth.

BEHAVIOUR The female wolf fish lays some of the largest eggs of all fish – about 5–6 mm (0.2–0.25 in) in diameter. The eggs are laid in a nest on the seabed, where the male guards them until the young are large enough to look after themselves.

01027346210211626.83.9 0102734

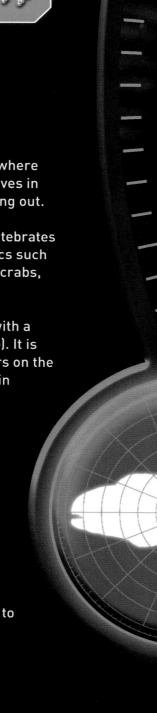

SIZE
1–1.5 m
3.3–5 FT

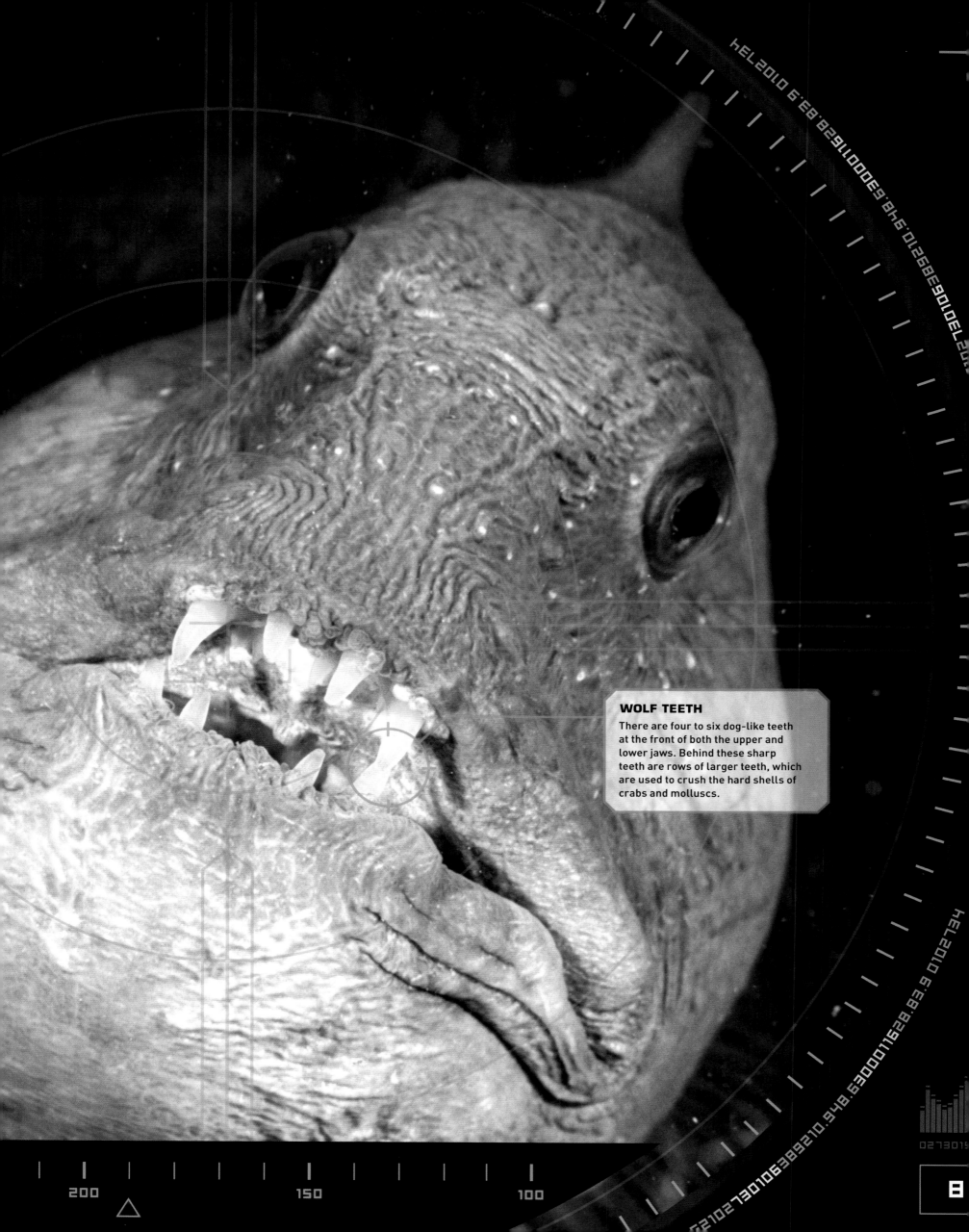

WOLF TEETH

There are four to six dog-like teeth at the front of both the upper and lower jaws. Behind these sharp teeth are rows of larger teeth, which are used to crush the hard shells of crabs and molluscs.

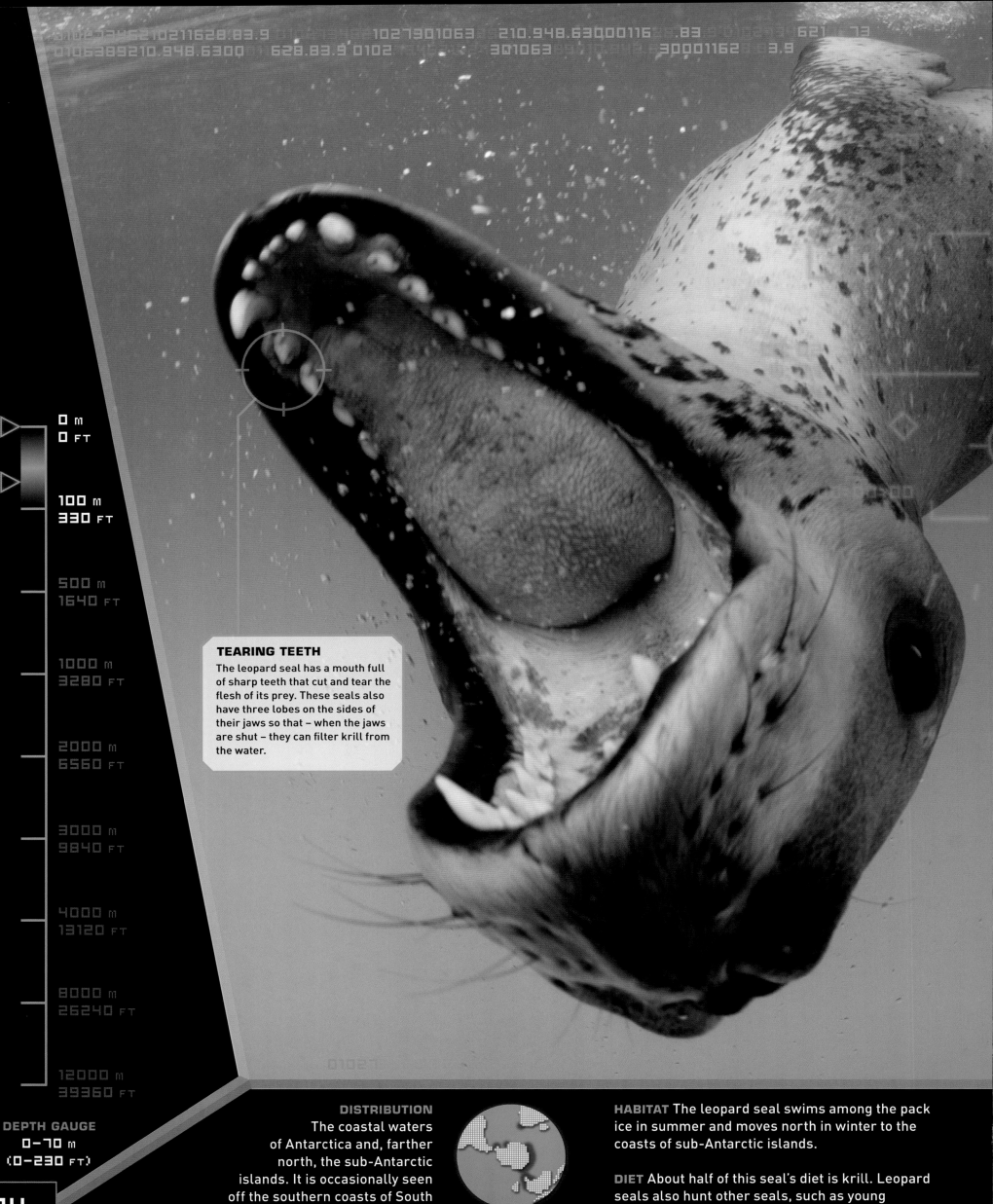

0 m
0 FT

100 m
330 FT

500 m
1640 FT

1000 m
3280 FT

2000 m
6560 FT

3000 m
9840 FT

4000 m
13120 FT

8000 m
26240 FT

12000 m
39360 FT

DEPTH GAUGE
0-70 m
(0-230 FT)

TEARING TEETH

The leopard seal has a mouth full
of sharp teeth that cut and tear the
flesh of its prey. These seals also
have three lobes on the sides of
their jaws so that – when the jaws
are shut – they can filter krill from
the water.

DISTRIBUTION
The coastal waters
of Antarctica and, farther
north, the sub-Antarctic
islands. It is occasionally seen
off the southern coasts of South
America, Australia and New Zealand.

Antarctic

HABITAT The leopard seal swims among the pack
ice in summer and moves north in winter to the
coasts of sub-Antarctic islands.

DIET About half of this seal's diet is krill. Leopard
seals also hunt other seals, such as young
crabeaters, as well as penguins, fish and squid.

GREAT STRENGTH

Leopard seals are large and well muscled. They use their strength to terrorize penguin colonies. They lie in wait under an icy ledge, watching for penguins to dive into the water. They then catch the birds by their feet and beat them against the surface of the water to kill them and rip them up.

SIZE
2.5–3.3 m
8–11 FT

LEOPARD SEAL

LATIN NAME *Hydrurga leptonyx*

With its wide, smiling mouth full of sharp teeth, the leopard seal is one of the most feared predators of the Antarctic. The seal gets its name from the dark spots on its throat, like those of a leopard. It hunts alone in the murky Antarctic waters. Its streamlined shape and powerful front flippers push it through the water so it can chase fast-swimming prey such as penguins. This is the only seal to prey on other seals.

BEHAVIOUR

When she gives birth, the female pulls herself onto the ice. Here she digs a hole in which to give birth to a single pup. The pup weighs about 30 kg (65 lb) and is about 1.6 m (5 ft) long. The female stays with her pup for a month, until it is large enough to fend for itself.

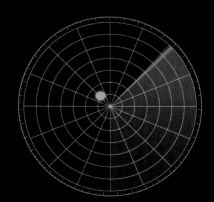

RANGEFINDER 10 cm (4 in)

DESCRIPTION The seal has a muscular body with a large head and long neck. Its coat is dark grey on the back and light grey on the underside, with dark spots on the white throat. Females are larger than males. Leopard seal pups have a dark stripe along the back and black spots on the underside.

RELATIVES The leopard seal is an earless seal belonging to the family of true seals, Phocidae. These seals are more suited to moving through water than on the land, as they have a streamlined but bulky body. Usually their front flippers are used to steer, but the leopard seal uses these flippers for propulsion.

brown speckles and lacks a covering of scales. This fish has transparent blood as it has no red blood cells. The ice fish has a long head with a narrow snout, and a large mouth and eyes. Its dorsal fin extends down the back, with large spines at the front.

NO HAEMOGLOBIN

The ice fish is the only vertebrate animal without haemoglobin in its blood. Haemoglobin transports oxygen from the lungs or gills to the rest of the body. But large red blood cells packed with haemoglobin make blood flow very slowly in cold water. Since oxygen can pass through the ice fish's skin, there is no need to have red blood cells.

0 m
0 FT

100 m
330 FT

500 m
1640 FT

1000 m
3280 FT

2000 m
6560 FT

3000 m
9840 FT

4000 m
13120 FT

8000 m
26240 FT

12000 m
39360 FT

DEPTH GAUGE
50–800 m
(165–2620 FT)

SIZE
30–75 CM
1–2.5 FT

DISTRIBUTION
Mostly in the Southern Ocean around Antarctica, but some are found around New Zealand and South America.

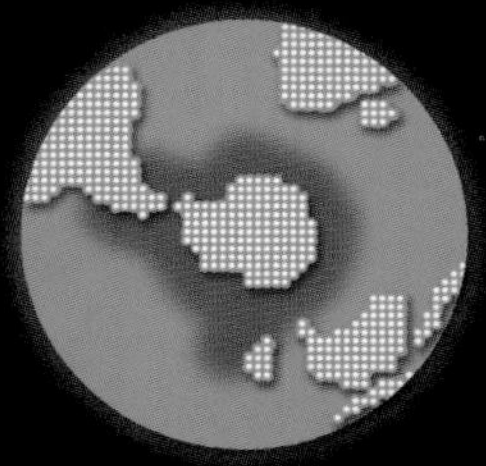

HABITAT Ice fish usually inhabit shallow coastal waters, where the water temperature is between 4°C (39°F) and -2°C (28°F). But a few have been found in much deeper water.

DIET The ice fish feeds on krill, copepods (small crustaceans) and fish.

CROCODILE ICE FISH

LATIN NAME Channichthyidae family

The crocodile, or white-blooded, ice fish are unique fish that are found in the waters around Antarctica. They are well adapted to living in the icy waters, where temperatures are around freezing point. They have their own antifreeze, which circulates around the body in the blood, stopping ice crystals from forming in the tissues.

NO SCALES
The ice fish gets its oxygen from the cold water, which is very well oxygenated. There are no scales covering the fish's body, so oxygen can pass straight through the skin as well as through the gills.

BEHAVIOUR The ice fish is slow to mature, so it does not spawn (produce eggs) before it is four to eight years of age. Spawning takes place in autumn and winter, when the female lays between 10,000 and 20,000 large, yolky eggs. The eggs hatch two months later.

SPECIES Crocodile ice fish belong to the family Channichthyidae. There are 24 species, most of which are found in the icy Antarctic waters. All these fish have a spiny dorsal fin and lack red blood cells with haemoglobin. There are over 120 species of ice fish in total.

RANGEFINDER 1 m (3 ft)

KRILL

LATIN NAME *Euphasia* sp.

Krill may be small but they are vital to the health of the world's oceans. Krill feed on tiny plant plankton floating in the sea, and then in turn are eaten by other animals, such as whales, fish, seals and birds. If the numbers of krill fall, the other animals cannot find enough food. This is happening now, especially in the Southern Ocean: Antarctic krill feed on algae that live under the ice, but as the ice melts, there are fewer algae.

0 m
0 FT

100 m
330 FT

500 m
1640 FT

1000 m
3280 FT

2000 m
6560 FT

3000 m
9840 FT

4000 m
13120 FT

8000 m
26240 FT

12000 m
39360 FT

DEPTH GAUGE
1–100 m
(3–330 FT)

DISTRIBUTION In all the world's oceans, in the upper 100 m (330 ft) of water. They are particularly abundant in the Southern Ocean.

HABITAT During the day krill are found at a depth of 100 m (330 ft), where it is dark, but during the night they go to the surface to feed.

DIET Krill are herbivores and feed on the plant plankton floating in the ocean and algae found on ice.

DESCRIPTION They range in size, but most krill are about 5–6 cm (2–2.4 in) long and weigh just a gram (0.04 oz). They look like miniature shrimps, with a body that is squashed from side to side. They look pink because their shell is transparent, and they have large black eyes. Krill have three pairs of feeding legs and five pairs of swimming legs.

SPECIES Krill are crustaceans, invertebrate animals that have a shell and jointed legs. There are about 85 species of krill, of which the most important is the Antarctic krill.

BEHAVIOUR A group of krill is called a swarm. At certain times of year, krill gather in huge swarms. These swarms are so dense that there are more than 60,000 individuals in 1 m³ (35 ft³) of water. Some swarms are so large that they can be seen by space station astronauts.

SWIMMING LEGS

A krill has five pairs of swimming legs. It uses these legs to swim down to deeper depths during the day and then drift upwards again at night to feed. Krill swim at a few centimetres per second. When danger threatens, they can speed backwards through the water by flicking their tail.

SIZE
1–15 CM
0.4–6 IN

FEEDING LEGS

Krill are filter feeders. They have long, comb-like feeding legs attached to their head-thorax. The combs sieve tiny particles of food from the water and then push them into the mouth.

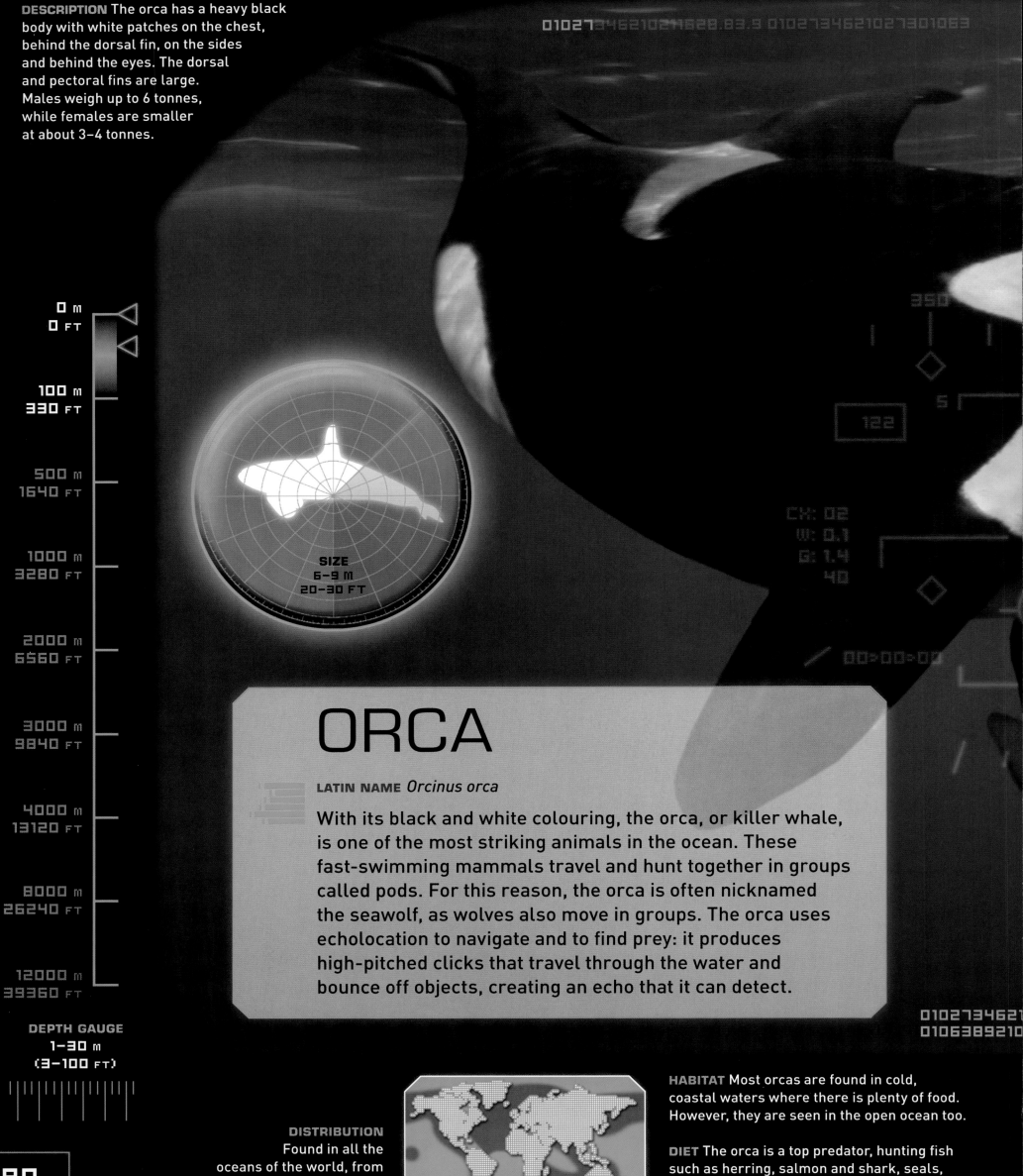

DESCRIPTION The orca has a heavy black body with white patches on the chest, behind the dorsal fin, on the sides and behind the eyes. The dorsal and pectoral fins are large. Males weigh up to 6 tonnes, while females are smaller at about 3–4 tonnes.

0 m
0 FT

100 m
330 FT

500 m
1640 FT

1000 m
3280 FT

2000 m
6560 FT

3000 m
9840 FT

4000 m
13120 FT

8000 m
26240 FT

12000 m
39360 FT

DEPTH GAUGE
1–30 m
(3–100 FT)

SIZE
6–9 M
20–30 FT

ORCA

LATIN NAME *Orcinus orca*

With its black and white colouring, the orca, or killer whale, is one of the most striking animals in the ocean. These fast-swimming mammals travel and hunt together in groups called pods. For this reason, the orca is often nicknamed the seawolf, as wolves also move in groups. The orca uses echolocation to navigate and to find prey: it produces high-pitched clicks that travel through the water and bounce off objects, creating an echo that it can detect.

DISTRIBUTION
Found in all the oceans of the world, from the Poles to the Equator.

HABITAT Most orcas are found in cold, coastal waters where there is plenty of food. However, they are seen in the open ocean too.

DIET The orca is a top predator, hunting fish such as herring, salmon and shark, seals, penguins and even other whales.

ACROBATIC ORCA

Orcas are excellent swimmers, with a streamlined shape that slips through the water. They can leap completely out of the sea and land back on the surface with a loud splash. At other times, they poke their head out of the water to have a good look around. This is called spy-hopping.

ORCA CALF

As soon as her calf is born, the female orca pushes it to the surface to take its first breath of air. Most calves stay with their mother for their entire lives. The orca is a very vocal whale and produces a range of sounds, such as clicks and whistles. A pod has its own particular sounds so that members can identify each other, and the females teach their calves the different sounds.

BEHAVIOUR Orcas in different regions prey on particular species and develop clever hunting strategies. Orcas along the coast of South America feed on sea lions. They can be seen riding the waves right up the beach, where they catch the seals in shallow water, but they have to be careful that they do not get stranded.

RELATIVES The orca is the largest of the 35 species of dolphin. Dolphins are toothed whales and are closely related to beluga and minke whales. Biologists studying the behaviour and diet of orcas think that there could be as many as five different orca subspecies.

RANGEFINDER 10 m (33 FT)

DESCRIPTION The male walrus weighs up to an incredible 2000 kg (4400 lb). Females weigh up to about 1300 kg (2900 lb). The walrus has a thick skin that is bald, apart from the bristles around the snout. The skin gets paler with age. There is a layer of blubber under the skin to insulate the body in cold water. Both sexes have tusks.

0 m	0 FT
100 m	330 FT
500 m	1640 FT
1000 m	3280 FT
2000 m	6560 FT
3000 m	9840 FT
4000 m	13120 FT
8000 m	26240 FT
12000 m	39360 FT

DEPTH GAUGE
0–80 m
(0–260 FT)

WALRUS

LATIN NAME *Odobenus rosmarus*

The walrus is an odd-looking marine mammal with a pair of long tusks, a beard of bristles, and a thick, wrinkly skin. It is found in the Arctic, where it lives in groups called herds. The walrus, like the related seals, is adapted to life in the water and has flippers. This creature is clumsy on land, but once in the water it makes a graceful swimmer. It swims like a seal, using its body rather than its flippers.

HUGE TUSKS
The longest tusks reach up to 1 m (3.3 ft) in length and may weigh more than 5 kg (11 lb). The males use their tusks for display and occasionally for fighting. Walruses use their tusks mostly for digging holes in the ice and hauling themselves onto the ice.

DISTRIBUTION
Found in the northern Atlantic and Pacific Oceans around the Arctic and Subarctic. Walruses move south in winter as the ice expands.

Arctic

HABITAT Walruses live in shallow coastal waters and are often seen on beaches and sea ice.

DIET The walrus eats animals living on the seabed, including shrimp, crabs, corals and sea cucumbers, but its favourite food is molluscs, especially mussels and clams.

BRISTLY WHISKERS

The snout is covered in long bristles or whiskers, each up to 30 cm (12 in) long. There are as many as 700 bristles arranged in rows. Each bristle is incredibly sensitive and they are used to find food in the mud of the seabed.

SIZE
2–3 M
6.5–10 FT

BEHAVIOUR During the breeding season, the males fight each other to win a group of females. Males stab at each other with their tusks. Fortunately, fights are over quickly and the loser moves away. The winner usually guards his group for a few days before he is replaced by another male.

SUBSPECIES There is only one species of walrus. The species is divided into three subspecies, which live in different regions. The two main subspecies are the Atlantic and Pacific walruses, while a third subspecies is found in the Laptev Sea off northeastern Russia.

RANGEFINDER 10 cm (4 in)

POLAR BEAR

LATIN NAME *Ursus maritimus*

The polar bear is the largest meat-eater found on land. This distinctive white bear is seen across the Arctic Ocean. It is a patient hunter, sitting by a hole in the sea ice, ready to grab a seal when it pops its head out of the water to breathe. Polar bears also stalk seals while they rest on the ice, creeping up on them very slowly. This bear is a scavenger too, relying on its remarkable sense of smell to find rotting carcasses on the ice.

0 m
0 FT

100 m
330 FT

500 m
1640 FT

1000 m
3280 FT

2000 m
6560 FT

3000 m
9840 FT

4000 m
13120 FT

8000 m
26240 FT

12000 m
39360 FT

DEPTH GAUGE
0–5 m
(0–16 FT)

Arctic

DISTRIBUTION Found across the Arctic, as far north as the North Pole and as far south as southern Greenland and Iceland.

HABITAT Polar bears live on the sea ice, hunting near the edge where it is thinnest. They have been seen in open Arctic waters 320 km (200 miles) from land.

DIET These bears eat mostly ringed seal, as well as small whales, fish, birds and rotting flesh.

DESCRIPTION Polar bears are large, stocky animals with hind legs that are longer than the front legs. A large male can weigh as much as 650 kg (1430 lb). The polar bear's fur is white but the skin underneath is black. It has furry, snowshoe-like paws that do not slip on the ice.

RELATIVES There is a single species of polar bear. Its closest relatives are the black and brown bears, which belong to the same genus, *Ursus*. Bears are mammals and are grouped with the other carnivores (meat-eaters), such as lions, tigers and wolves.

BEHAVIOUR During the autumn, the pregnant female moves inland to make a den in the snow. She stays in her den, not eating or drinking, all through the winter. She gives birth to two or three cubs in December. The female and her cubs emerge from the den three months later.

GOOD SWIMMER

The polar bear is a strong swimmer, travelling at speeds of up to 10 kph (6 mph). It can swim over distances of up to 100 km (60 miles) when the sea ice starts to break up. It swims using its large front paws, and can dive under the water for up to two minutes. While under water, it closes its nostrils and holds its ears flat against its head.

SIZE
2–3 m
6.5–10 FT

STAYING WARM

Polar bears stay warm in the icy water because they have a thick layer of blubber underneath their skin. It is up to 10 cm (4 in) thick and stops heat escaping from the body. This insulation is so good that polar bears can overheat. On a warm day they may dive into the water to cool down.

ON THE HIGH SEAS

Leaving the land far behind, we travel out into the vast wilderness that is the open ocean, before starting our eerie descent into the darkest depths.

The coast is fringed with shallow water lying over the continental shelf, but farther out the ocean gets deeper, much deeper. In places the seabed slopes down to depths of more than 6000 m (20000 ft). The oceans are so deep that whole mountain chains are under water. Take away the water and you would be left with an incredible landscape of towering mountains and plunging trenches that would better anything found on land.

On a sunny day, glimmers of sunlight can reach to depths of 200 m (650 ft). This marks the bottom of the sunlight zone, where we can find a wealth of species, from stinging jellyfish to terrifying predators such as the great white shark. Below the sunlight zone, the deep ocean forms the largest habitat on Earth. Amazingly, this is the habitat that we know the least about. Scientists once believed that little could survive in the deep, but they were wrong. There is an extraordinary variety of animals here, and almost every journey into the deep discovers an animal new to science.

OPEN OCEAN

Far away from land, our submersible journeys through the upper, warmer layers of water, searching for predators and their prey.

The open ocean is vast, in places stretching for thousands of kilometres with no land in sight. The surface layer of water is warmed by sunlight. Warm water is less dense than cold water, so it sits on top of the layers of cold water. The surface water is mixed by the waves and currents that bring nutrients up from the deep.

Invisible to the human eye is a wealth of tiny plant and animal life floating in the water. This is the plankton and it forms the basis for the ocean food chains. The plankton is eaten by fish, squid and even by the largest sharks and whales – including the biggest animal on Earth, the blue whale. Without plankton there would be little life in the oceans. By day, the surface layer of water seems empty of life. Most of the animals, plankton included, swim down to the safety of the dark water below, only returning to the surface at night.

02734621027301063B9210.94B.6300011628.B3.9 01027346389210.
B.6302102730106389210.948.630

GIANT SHOALS

Huge shoals of fish can be found in the open ocean. These food sources attract the large ocean predators, such as the blue shark, dolphins and even the extraordinary-looking ocean sunfish.

CROWN JELLYFISH

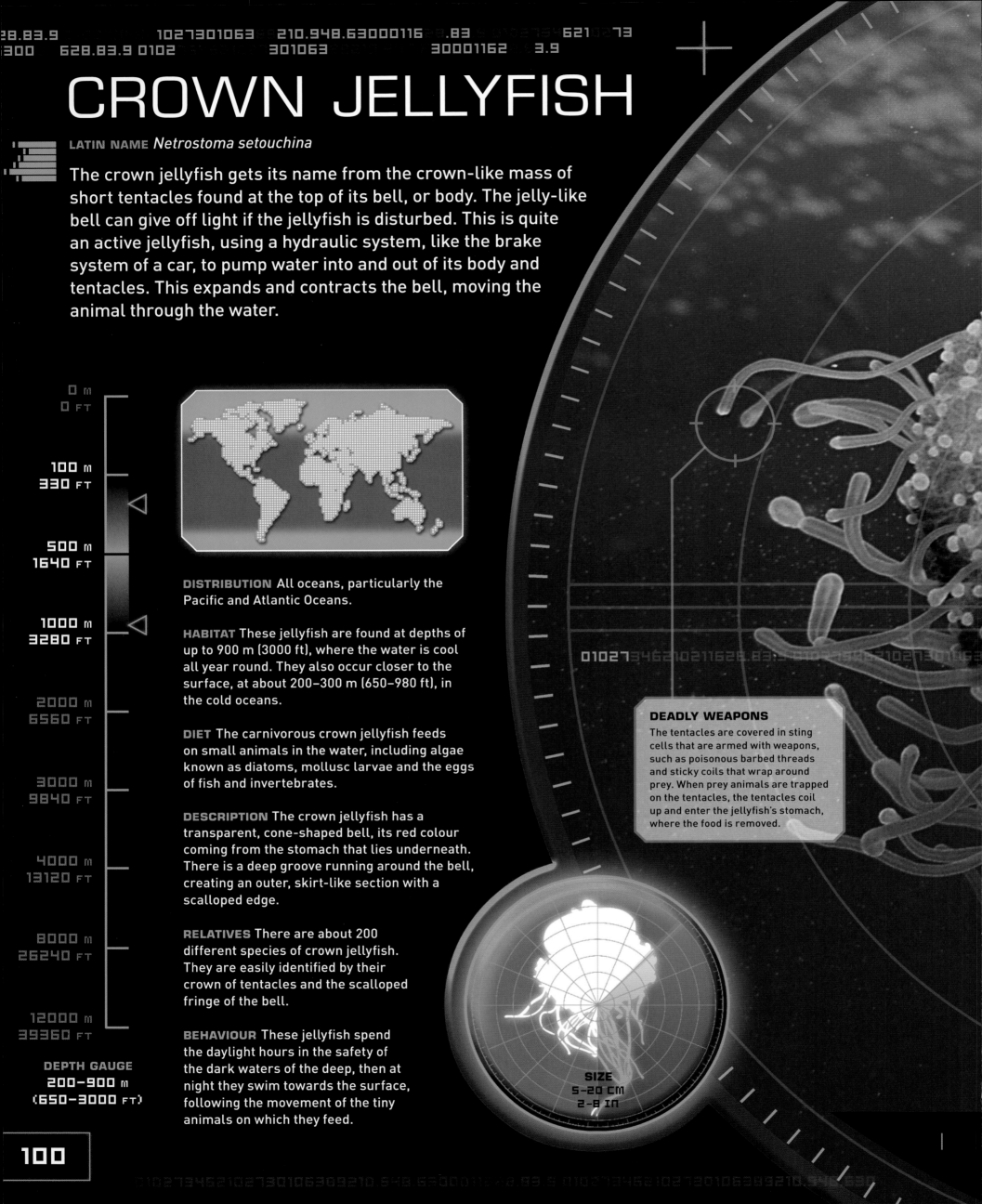

LATIN NAME *Netrostoma setouchina*

The crown jellyfish gets its name from the crown-like mass of short tentacles found at the top of its bell, or body. The jelly-like bell can give off light if the jellyfish is disturbed. This is quite an active jellyfish, using a hydraulic system, like the brake system of a car, to pump water into and out of its body and tentacles. This expands and contracts the bell, moving the animal through the water.

0 m
0 FT

100 m
330 FT

500 m
1640 FT

1000 m
3280 FT

2000 m
6560 FT

3000 m
9840 FT

4000 m
13120 FT

8000 m
26240 FT

12000 m
39360 FT

DEPTH GAUGE
200–900 m
(650–3000 FT)

DISTRIBUTION All oceans, particularly the Pacific and Atlantic Oceans.

HABITAT These jellyfish are found at depths of up to 900 m (3000 ft), where the water is cool all year round. They also occur closer to the surface, at about 200–300 m (650–980 ft), in the cold oceans.

DIET The carnivorous crown jellyfish feeds on small animals in the water, including algae known as diatoms, mollusc larvae and the eggs of fish and invertebrates.

DESCRIPTION The crown jellyfish has a transparent, cone-shaped bell, its red colour coming from the stomach that lies underneath. There is a deep groove running around the bell, creating an outer, skirt-like section with a scalloped edge.

RELATIVES There are about 200 different species of crown jellyfish. They are easily identified by their crown of tentacles and the scalloped fringe of the bell.

BEHAVIOUR These jellyfish spend the daylight hours in the safety of the dark waters of the deep, then at night they swim towards the surface, following the movement of the tiny animals on which they feed.

DEADLY WEAPONS
The tentacles are covered in sting cells that are armed with weapons, such as poisonous barbed threads and sticky coils that wrap around prey. When prey animals are trapped on the tentacles, the tentacles coil up and enter the jellyfish's stomach, where the food is removed.

SIZE
5–20 CM
2–8 IN

FEEDING TACTICS

These jellyfish have a particular feeding behaviour. They hold their tentacles alongside their bell while swimming downwards for about 10 m (30 ft). Then they drift upwards, trailing their tentacles behind them, catching small animals in the water.

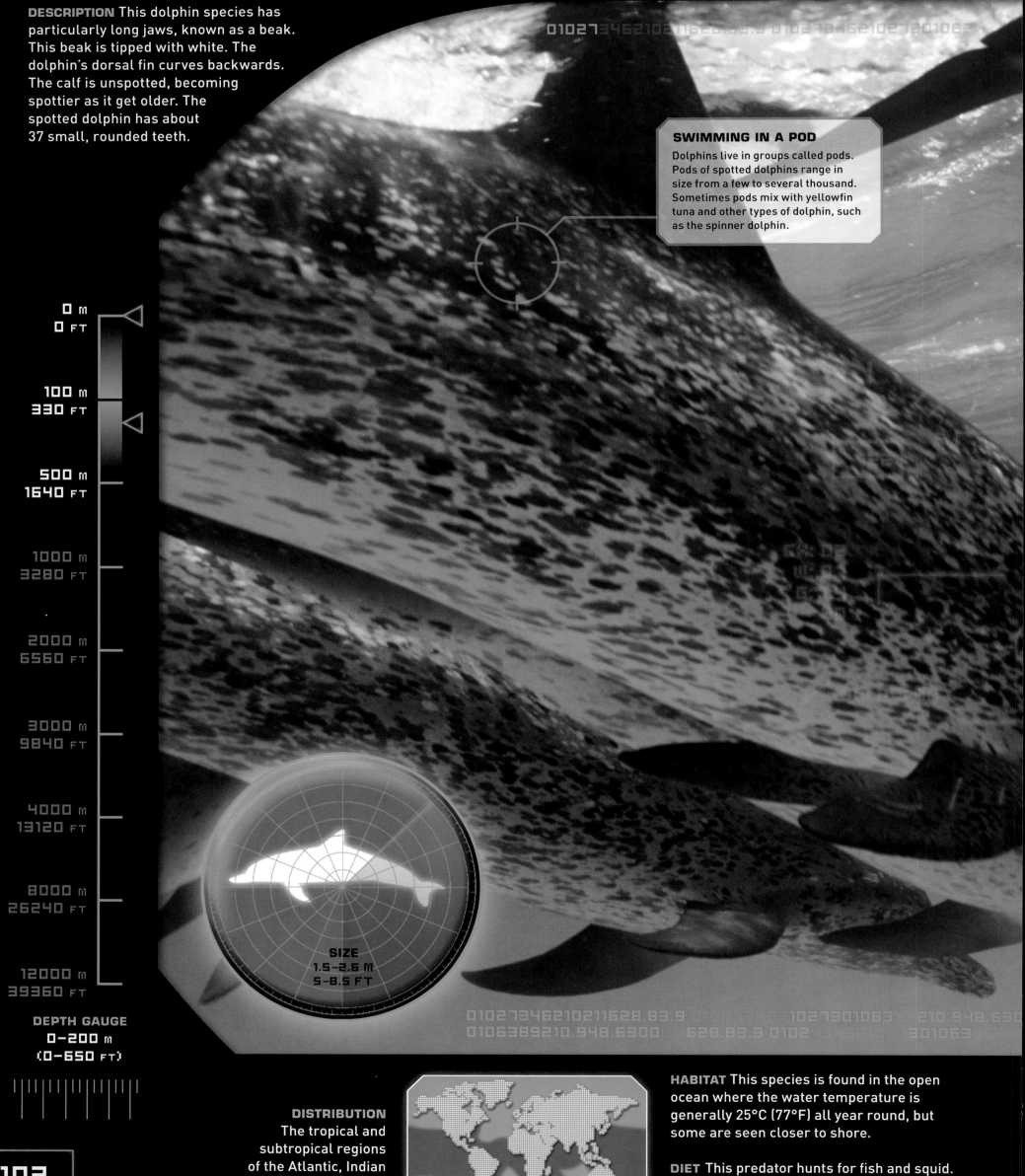

DESCRIPTION This dolphin species has particularly long jaws, known as a beak. This beak is tipped with white. The dolphin's dorsal fin curves backwards. The calf is unspotted, becoming spottier as it get older. The spotted dolphin has about 37 small, rounded teeth.

SWIMMING IN A POD
Dolphins live in groups called pods. Pods of spotted dolphins range in size from a few to several thousand. Sometimes pods mix with yellowfin tuna and other types of dolphin, such as the spinner dolphin.

0 m
0 FT

100 m
330 FT

500 m
1640 FT

1000 m
3280 FT

2000 m
6560 FT

3000 m
9840 FT

4000 m
13120 FT

8000 m
26240 FT

12000 m
39360 FT

DEPTH GAUGE
0–200 m
(0–650 FT)

SIZE
1.5–2.6 M
5–8.5 FT

DISTRIBUTION
The tropical and subtropical regions of the Atlantic, Indian and Pacific Oceans.

HABITAT This species is found in the open ocean where the water temperature is generally 25°C (77°F) all year round, but some are seen closer to shore.

DIET This predator hunts for fish and squid. It is particularly active at night.

SPOTTED DOLPHIN

LATIN NAME *Stenella attenuata*

The spotted dolphin, named after the spots on its body, is a fast-swimming species. Its sleek shape allows it to slip through the water at up to 28 kph (16 mph). It is acrobatic too, often seen leaping completely out of the water. This dolphin has very little blubber under its skin, which means it lacks reserves of fat for storing energy. It prefers to swim in warm water, where it feeds on energy-rich foods such as fish.

BULGING MELON

The bulging part of the dolphin's forehead is the melon, which is used in echolocation. Dolphins produce a series of clicks and the melon focuses these sounds into a beam that travels through the water. The sounds bounce off objects and produce an echo. The dolphin works out the distance of the object from the time it takes the echo to return.

BEHAVIOUR Some groups of spotted dolphins migrate each year. For example, in the Pacific the dolphins swim north in summer and return in winter. This dolphin often swims closely with yellowfin tuna and may be accidentally caught and killed by tuna fishermen. Today many fishermen practise dolphin-safe fishing.

0273019210

SUBSPECIES The dolphin is a type of toothed whale. The spotted dolphin is quite variable in appearance, especially in its length and the number of spots: three subspecies are thought to exist. Some biologists recognize a Hawaiian, an offshore and a coastal subspecies.

RANGEFINDER 1 m (3 FT)

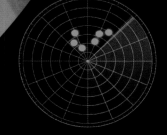

SARGASSUM FROGFISH

LATIN NAME *Histrio histrio*

This fish is a master of disguise. Its body looks like a piece of floating seaweed. Most frogfish are found on coral reefs and sandy seabeds, but the sargassum frogfish lives in floating rafts of sargassum algae, a type of seaweed. It has a dorsal spine with a fleshy tip, which when wiggled looks like a small fish. This attracts other fish, which it then eats.

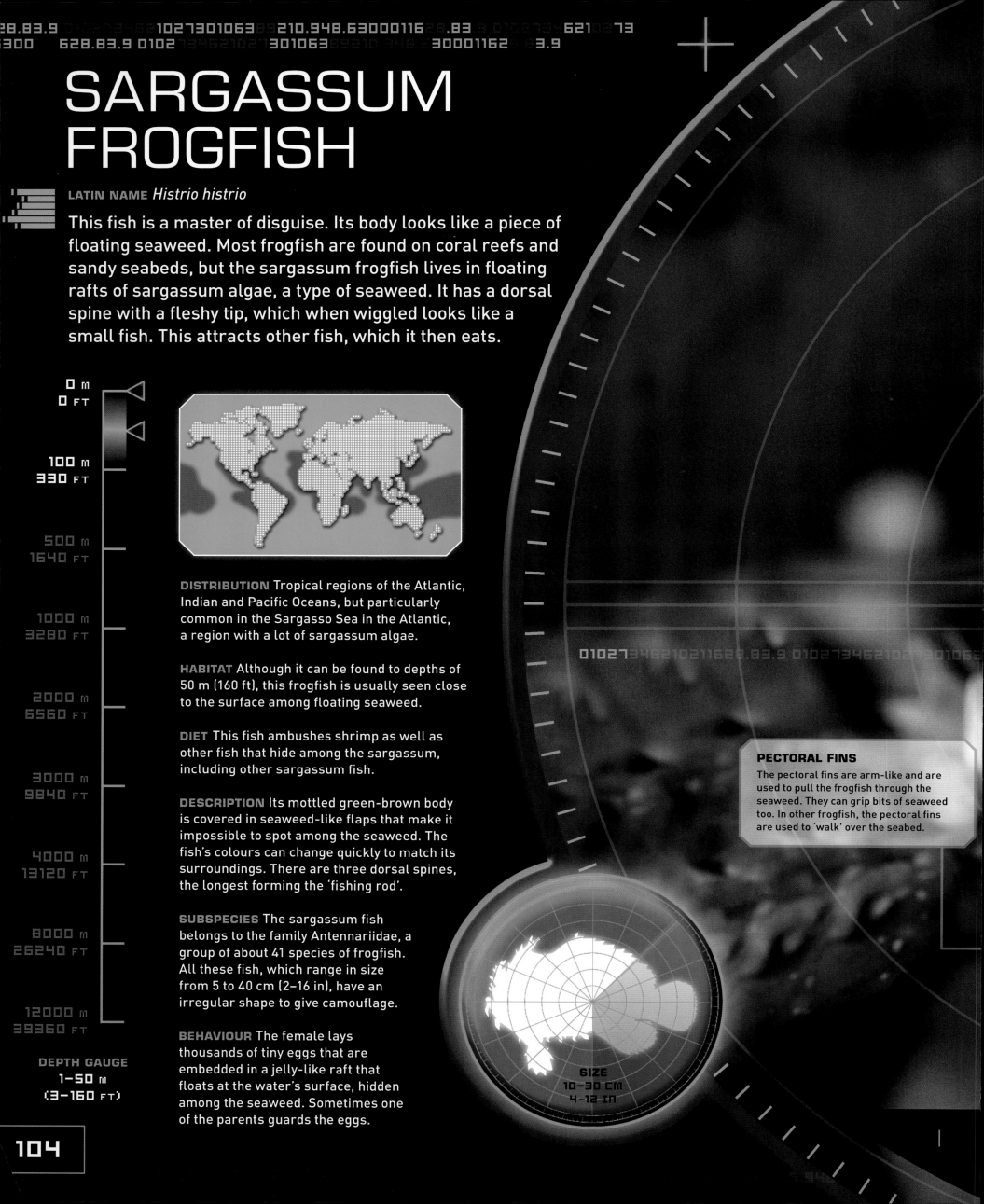

0 m
0 FT

100 m
330 FT

500 m
1640 FT

1000 m
3280 FT

2000 m
6560 FT

3000 m
9840 FT

4000 m
13120 FT

8000 m
26240 FT

12000 m
39360 FT

DEPTH GAUGE
1–50 m
(3–160 FT)

DISTRIBUTION Tropical regions of the Atlantic, Indian and Pacific Oceans, but particularly common in the Sargasso Sea in the Atlantic, a region with a lot of sargassum algae.

HABITAT Although it can be found to depths of 50 m (160 ft), this frogfish is usually seen close to the surface among floating seaweed.

DIET This fish ambushes shrimp as well as other fish that hide among the sargassum, including other sargassum fish.

DESCRIPTION Its mottled green-brown body is covered in seaweed-like flaps that make it impossible to spot among the seaweed. The fish's colours can change quickly to match its surroundings. There are three dorsal spines, the longest forming the 'fishing rod'.

SUBSPECIES The sargassum fish belongs to the family Antennariidae, a group of about 41 species of frogfish. All these fish, which range in size from 5 to 40 cm (2–16 in), have an irregular shape to give camouflage.

BEHAVIOUR The female lays thousands of tiny eggs that are embedded in a jelly-like raft that floats at the water's surface, hidden among the seaweed. Sometimes one of the parents guards the eggs.

PECTORAL FINS
The pectoral fins are arm-like and are used to pull the frogfish through the seaweed. They can grip bits of seaweed too. In other frogfish, the pectoral fins are used to 'walk' over the seabed.

SIZE
10–30 CM
4–12 IN

01027 3462102 11628 83.9 01027346210 301063

WIDE MOUTH
The jaw is hinged so that it can open really wide, while the stomach can stretch. This allows the fish to eat prey animals that are almost twice its own size, including other frogfish.

CROSS JELLY

LATIN NAME *Mitrocoma cellularia*

The cross jelly is a jellyfish-like animal that lives among the plankton. Plankton is a vital part of the ocean food chain as the small drifting plants and animals of which it consists are eaten by animals such as small fish, which in turn are eaten by larger predators, such as dolphins and sharks. The cross jelly is bioluminescent, which means it can produce its own eerie green light that is seen around the edge of its bell.

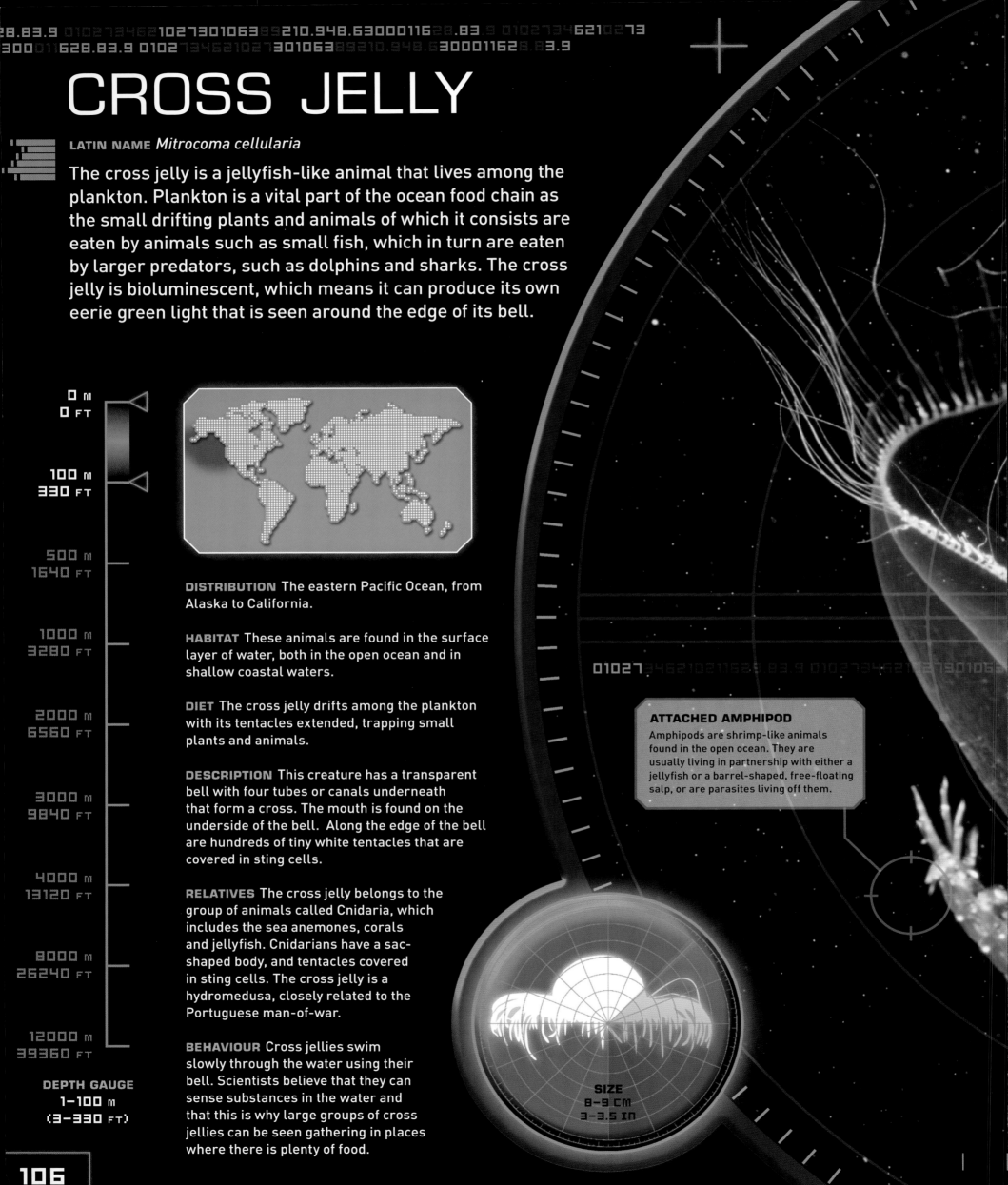

0 m
0 FT

100 m
330 FT

500 m
1640 FT

1000 m
3280 FT

2000 m
6560 FT

3000 m
9840 FT

4000 m
13120 FT

8000 m
26240 FT

12000 m
39360 FT

DEPTH GAUGE
1–100 m
(3–330 FT)

DISTRIBUTION The eastern Pacific Ocean, from Alaska to California.

HABITAT These animals are found in the surface layer of water, both in the open ocean and in shallow coastal waters.

DIET The cross jelly drifts among the plankton with its tentacles extended, trapping small plants and animals.

DESCRIPTION This creature has a transparent bell with four tubes or canals underneath that form a cross. The mouth is found on the underside of the bell. Along the edge of the bell are hundreds of tiny white tentacles that are covered in sting cells.

RELATIVES The cross jelly belongs to the group of animals called Cnidaria, which includes the sea anemones, corals and jellyfish. Cnidarians have a sac-shaped body, and tentacles covered in sting cells. The cross jelly is a hydromedusa, closely related to the Portuguese man-of-war.

BEHAVIOUR Cross jellies swim slowly through the water using their bell. Scientists believe that they can sense substances in the water and that this is why large groups of cross jellies can be seen gathering in places where there is plenty of food.

01027346210211628.83.9 010273462102730106

ATTACHED AMPHIPOD
Amphipods are shrimp-like animals found in the open ocean. They are usually living in partnership with either a jellyfish or a barrel-shaped, free-floating salp, or are parasites living off them.

SIZE
8–9 CM
3–3.5 IN

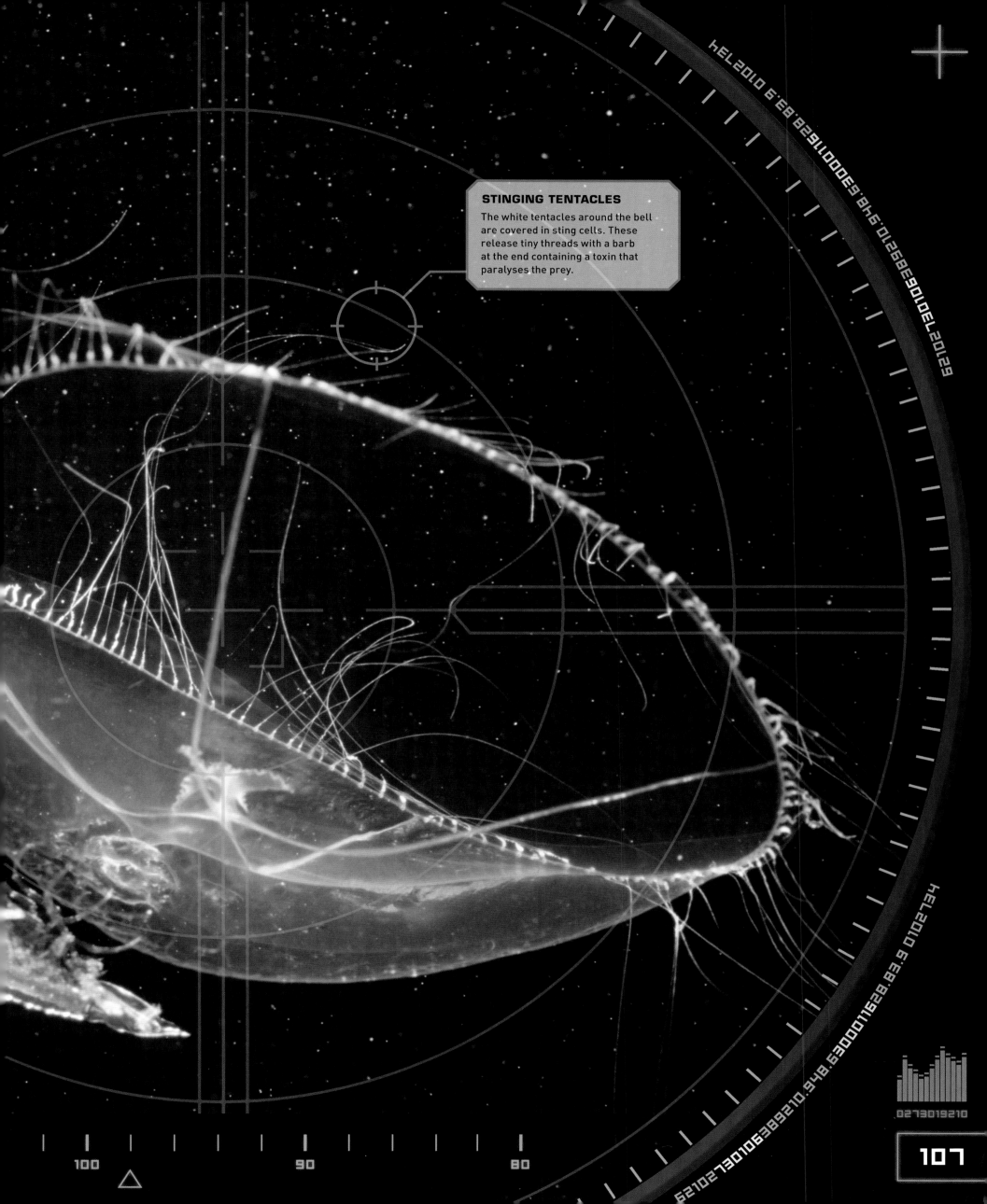

STINGING TENTACLES

The white tentacles around the bell are covered in sting cells. These release tiny threads with a barb at the end containing a toxin that paralyses the prey.

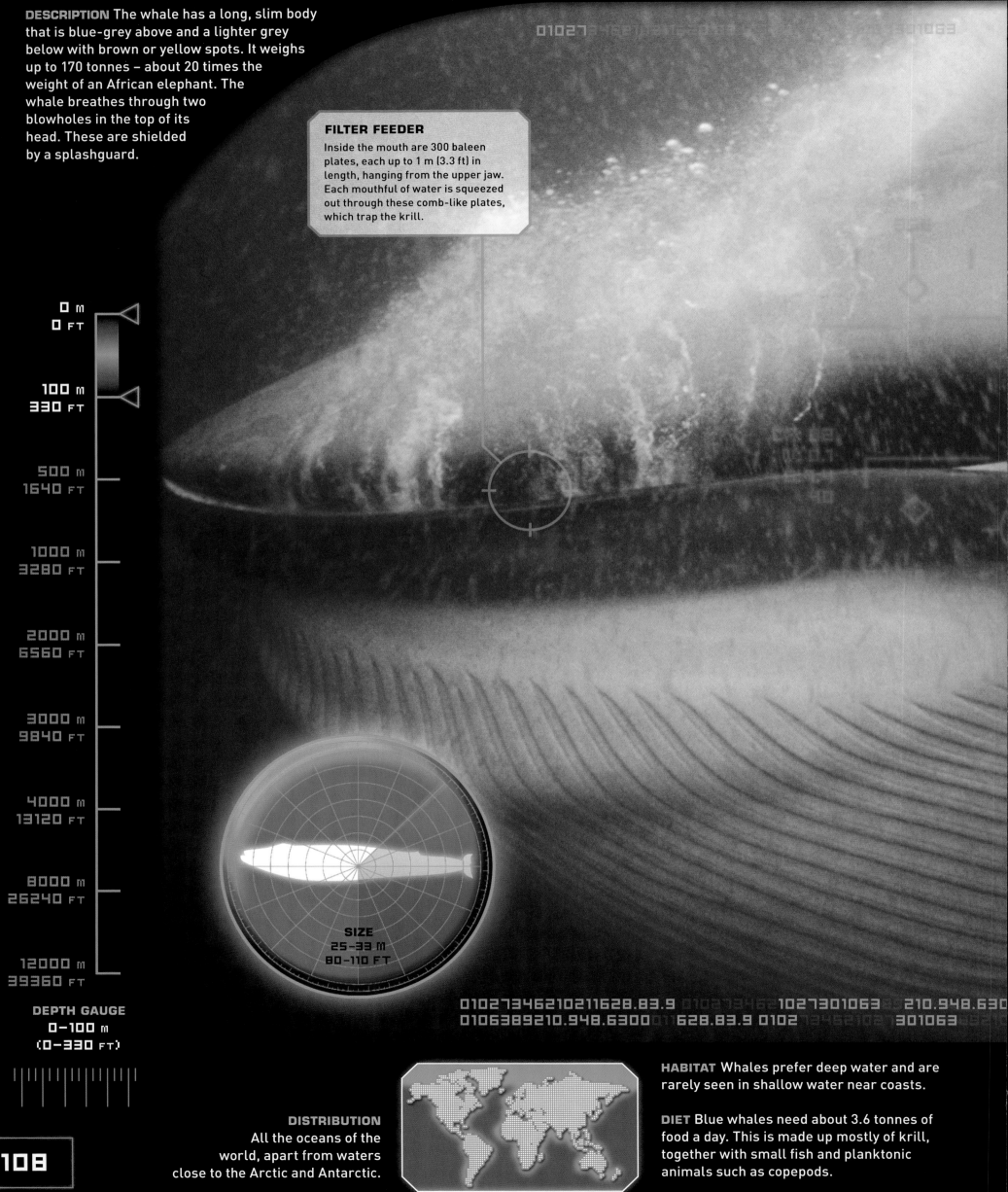

DESCRIPTION The whale has a long, slim body that is blue-grey above and a lighter grey below with brown or yellow spots. It weighs up to 170 tonnes – about 20 times the weight of an African elephant. The whale breathes through two blowholes in the top of its head. These are shielded by a splashguard.

FILTER FEEDER
Inside the mouth are 300 baleen plates, each up to 1 m (3.3 ft) in length, hanging from the upper jaw. Each mouthful of water is squeezed out through these comb-like plates, which trap the krill.

0 m
0 FT

100 m
330 FT

500 m
1640 FT

1000 m
3280 FT

2000 m
6560 FT

3000 m
9840 FT

4000 m
13120 FT

8000 m
26240 FT

12000 m
39360 FT

DEPTH GAUGE
0–100 m
(0–330 FT)

SIZE
25–33 M
80–110 FT

DISTRIBUTION
All the oceans of the world, apart from waters close to the Arctic and Antarctic.

HABITAT Whales prefer deep water and are rarely seen in shallow water near coasts.

DIET Blue whales need about 3.6 tonnes of food a day. This is made up mostly of krill, together with small fish and planktonic animals such as copepods.

106389210.948.63002734621027301063895210.94

BLUE WHALE

LATIN NAME *Balaenoptera musculus*

The blue whale is the largest animal on Earth and it may even be the largest animal ever to have lived. Everything about this whale is big – its heart is the size of a car, more than 6 tonnes of blood flow around its body, and even its tongue weighs 3 tonnes. Blue whales are either solitary or found in the company of one other whale. During the day they dive to depths of up to 100 m (330 ft) to search for krill. At night they feed on krill near the surface.

EXPANDING THROAT

There are up to 90 throat grooves running from the lower jaw over the chest. These expand when the whale swallows water, and then they help to push the water out after it has been filtered. The whale swallows up to 90 tonnes of water in one go.

26.83 010273 6210273
30001162 3.9

BEHAVIOUR Blue whales are the loudest animals on Earth. They can produce sounds that reach almost 190 decibels, far noisier than a jet engine. The sounds travel through the water for hundreds of kilometres and enable the whale to communicate with other whales and to find a mate.

0273019210

SUBSPECIES There are three subspecies of blue whale: the north blue whale of the North Atlantic and North Pacific Oceans, the southern blue whale of the Southern Ocean, and the pygmy blue whale of the Indian and Southern Pacific Oceans.

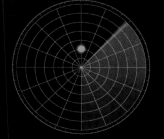

RANGEFINDER 8 m (26 FT)

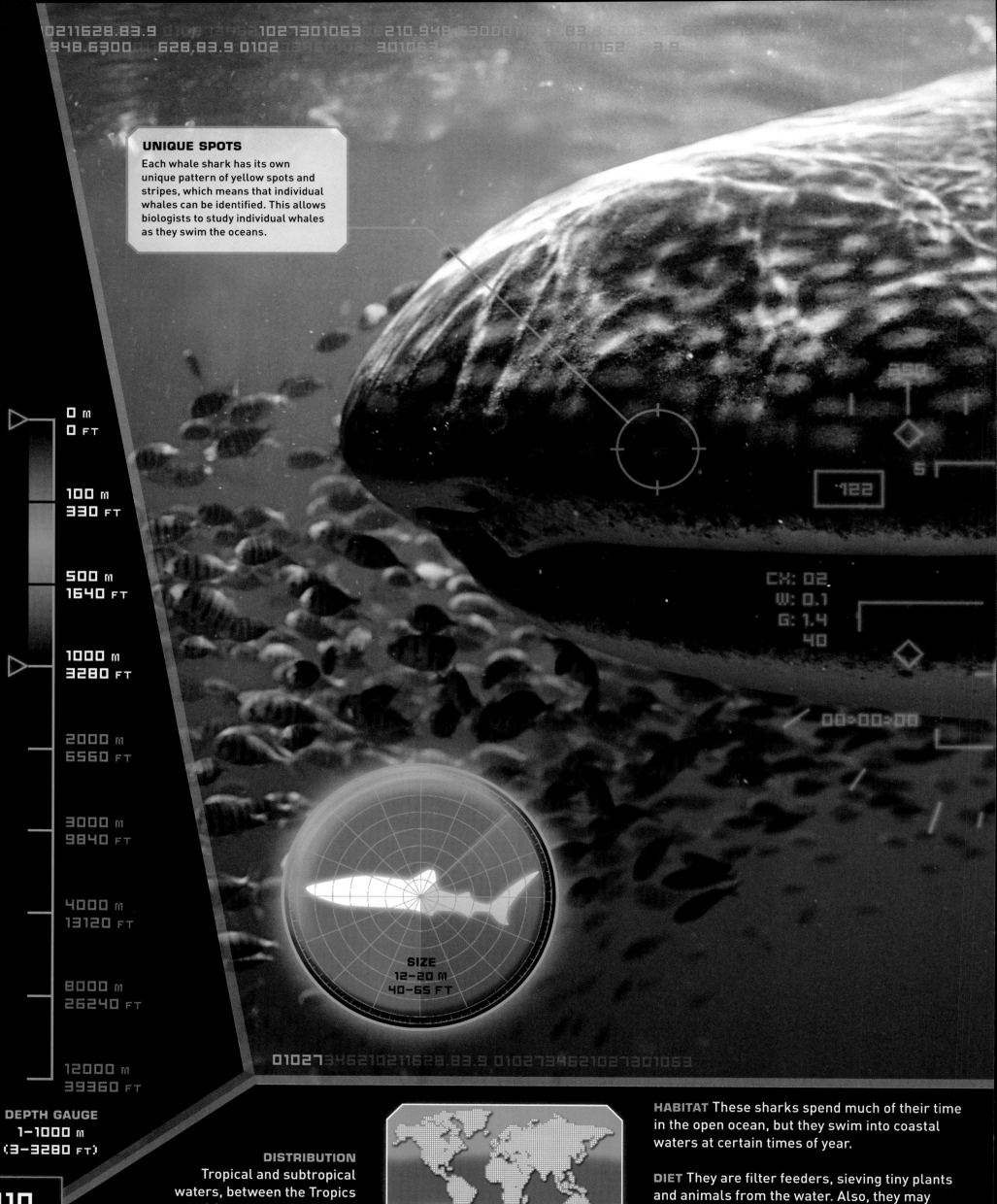

UNIQUE SPOTS
Each whale shark has its own unique pattern of yellow spots and stripes, which means that individual whales can be identified. This allows biologists to study individual whales as they swim the oceans.

0 m
0 FT

100 m
330 FT

500 m
1640 FT

1000 m
3280 FT

2000 m
6560 FT

3000 m
9840 FT

4000 m
13120 FT

8000 m
26240 FT

12000 m
39360 FT

CX: 02
W: 0.1
G: 1.4
40

00>00>00

SIZE
12–20 M
40–65 FT

DEPTH GAUGE
1–1000 m
(3–3280 FT)

DISTRIBUTION
Tropical and subtropical waters, between the Tropics of Cancer and Capricorn.

HABITAT These sharks spend much of their time in the open ocean, but they swim into coastal waters at certain times of year.

DIET They are filter feeders, sieving tiny plants and animals from the water. Also, they may swallow small squid, crustaceans and fish.

BIG MOUTH

The wide mouth is a massive 1.5 m (5 ft) across, with about 300 rows of tiny teeth on each jaw. However, whale sharks do not use their teeth to feed. Instead, they suck in mouthfuls of water that is squeezed through comb-like gill rakers that sieve the food from the water.

0027

00>06>22

WHALE SHARK

LATIN NAME *Rhincodon typus*

This is the gaping mouth of the largest fish in the ocean – the whale shark. The fish may be huge but it is harmless, feeding on plankton floating in the water. Its skin is a massive 10 cm (4 in) thick, the thickest of any animal. Whale sharks are mostly solitary, swimming slowly through the open ocean. The life cycle of this giant is still to be studied. Biologists know that the females give birth to live young, but little else.

BEHAVIOUR

Some whale sharks migrate across the oceans each year. One individual travelled more than 13,000 km (8000 miles) in three years. However, other whale sharks stay in the same region all year round. The journeys are thought to coincide with the annual spawning of fish and plankton.

RANGEFINDER **2** m (6 FT)

DESCRIPTION The whale shark has a broad, flat head with a wide mouth and tiny eyes. Most of these sharks weigh up to 14 tonnes, but 34 tonnes has been recorded. The body is grey with a white underside and is marked with spots and stripes. The tail fin is crescent-shaped.

RELATIVES The whale shark belongs to the carpet shark order. Most carpet sharks, such as the wobbegong, are small and live on the seabed. The whale shark is the only species in its family, Rhincodontidae. It is not related to the whales: it gets it name from its size and its method of filter-feeding like a baleen whale.

0273019210

PORTUGUESE MAN-OF-WAR

LATIN NAME *Physalia physalis*

The odd name of this animal comes from its triangular float, or sail, that rises above the water. It was likened to the sail of an old Portuguese warship. Even stranger is the fact that the Portuguese man-of-war is not one whole animal, but is made up of a colony of tiny tentacled animals called polyps which are attached to each other.

0 m
0 FT

100 m
330 FT

500 m
1640 FT

1000 m
3280 FT

DISTRIBUTION The warm waters of the Atlantic, Indian and Pacific Oceans.

HABITAT They drift on the open ocean, carried by currents and winds. Sometimes they are carried into shallow coastal waters.

2000 m
6560 FT

DIET Man-of-wars feed on small animals, such as plankton, crustaceans and fish.

DESCRIPTION The gas-filled float is formed from a single polyp. It is blue to purple in colour and up to 15 cm (6 in) high. The tentacles can extend up to 50 m (165 ft) in length, although 1 m (3.3 ft) is normal. There are capturing tentacles covered in sting cells, feeding tentacles and reproductive tentacles.

3000 m
9840 FT

4000 m
13120 FT

RELATIVES The Portuguese man-of-war is a Cnidarian, related to the sea anemones and jellyfish. There are two closely related species of man-of-war: the Portuguese is widely distributed, but the other species, called the bluebottle, is found only in the Indo-Pacific Oceans.

8000 m
26240 FT

12000 m
39360 FT

DEPTH GAUGE
0 m
(0 FT)

BEHAVIOUR The float is filled with gas so it stays at the surface. However, if the man-of-war is attacked from above, it deflates its float so it sinks until the danger has passed.

01027 346210211628.83.9 01027346210 7301063

SAILING ALONG

The crest at the top of the float is raised like a sail to catch the wind. The float needs to stay moist, so every now and again the man-of-war rolls sideways to dip into the water.

SIZE
2–50 M
6.5–165 FT

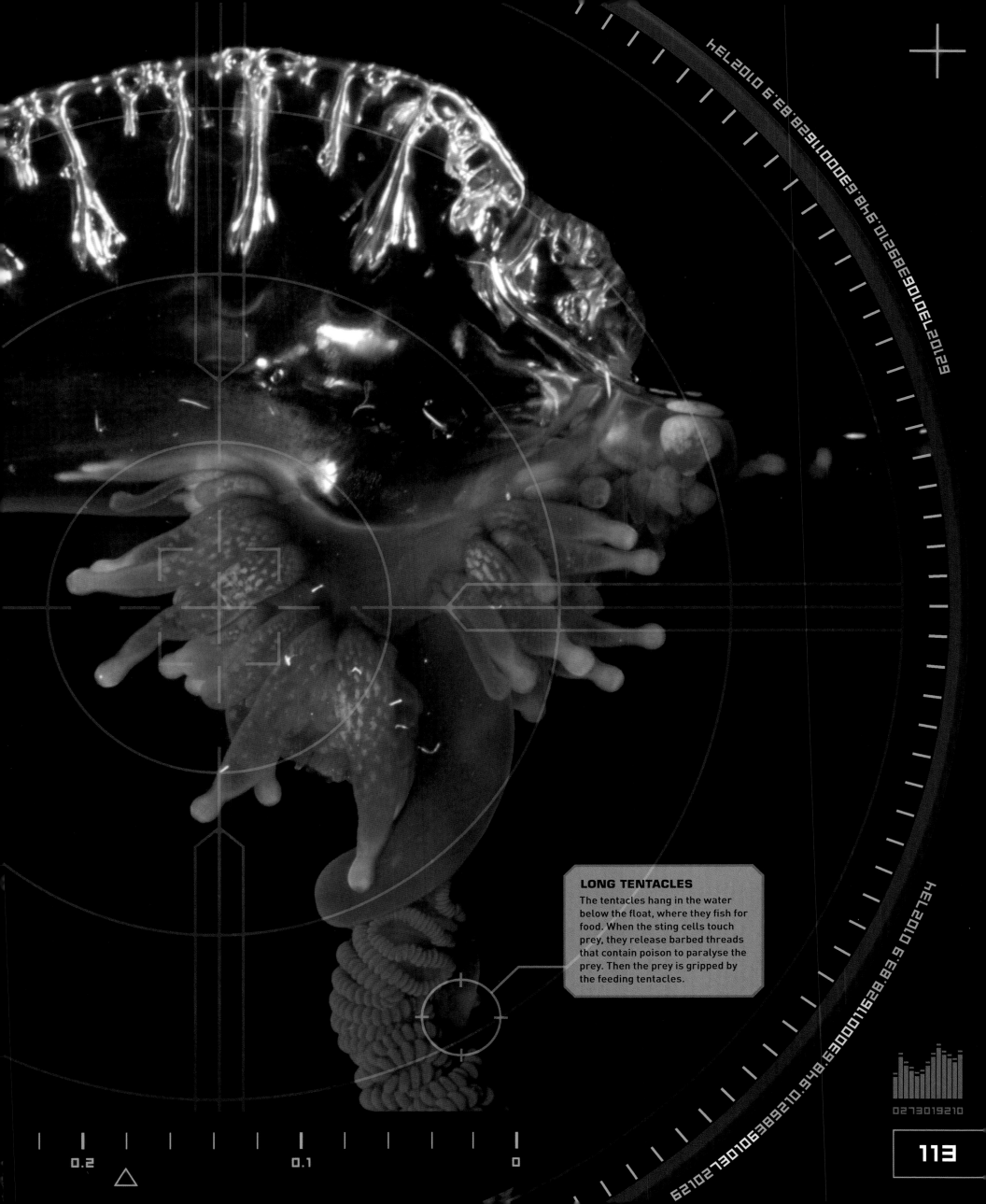

LONG TENTACLES

The tentacles hang in the water below the float, where they fish for food. When the sting cells touch prey, they release barbed threads that contain poison to paralyse the prey. Then the prey is gripped by the feeding tentacles.

0.2   0.1   0

DESCRIPTION This large fish weighs up to 3000 kg (6600 lb). Its upper body is grey and the underside is white. When seen from above, the shark blends with the dark water. When seen from below, it blends with the light-dappled water surface. The body is streamlined, helping the shark to power through the water at up to 24 kph (15 mph).

SENSING BLOOD
Scientists believe that the great white can detect tiny amounts of blood in the water from 5 km (3 miles) away. They taste the water using their nostrils.

JAWS
The great white has a mouth full of rows of jagged triangular teeth, each up to 7 cm (2.75 in) long. The teeth are constantly replaced. Once the shark has caught its prey, it shakes its head from side to side so that the teeth can rip through the flesh and tear off lumps.

0 m	0 FT
100 m	330 FT
500 m	1640 FT
1000 m	3280 FT
2000 m	6560 FT
3000 m	9840 FT
4000 m	13120 FT
8000 m	26240 FT
12000 m	39360 FT

DEPTH GAUGE
0–1300 m
(0–4260 FT)

DISTRIBUTION
The temperate oceans of the world, especially along the coasts of Australia, South Africa and California, USA.

HABITAT These sharks are found in cool to warm coastal waters.

DIET Great whites hunt seals, sea lions, small toothed whales and dolphins, turtles and fish, including rays. They may also scavenge on dead bodies floating in the water.

GREAT WHITE SHARK

LATIN NAME *Carcharodon carcharias*

Just the mention of the name 'great white shark' sends shivers down the spines of divers. With its mouth full of jagged teeth, it is not surprising that this fish is a much feared predator. It attacks from below, its dark upper body providing camouflage. Sometimes, it swims so quickly towards the surface that it leaps straight out of the water with its mouth full of struggling prey. It is responsible for one third of shark attacks on humans.

SIZE
4–7 M
13–23 FT

RANGEFINDER 1 m (3 FT)

BEHAVIOUR Unlike the majority of fish, the female great white keeps her eggs within her body. They hatch inside her and start to grow, but they receive no food from her. Instead, the babies feed on unhatched eggs and may even attack each other. Eventually, up to 14 pups are born, each about 1.5 m (5 ft) in length.

RELATIVES The great white belongs to the family of mackerel sharks, the fast-swimming sharks that include the mako and porbeagle sharks. The great white may be related to the extinct *Megalodon* shark, which grew to 18 m (60 ft) long and probably died out about 1.5 million years ago.

CEAN SUNFISH

NAME *Mola mola*

odd-looking fish has a round body, which is flattened
side to side, and a back end that looks as if it has been
ped off! The name sunfish comes from the fish's habit
ng on its side at the surface of the ocean, as if it is
athing. The sunfish suffers from skin parasites, so it
vs wrasse and other cleaner fish to remove them.
heavy fish can weigh a massive 2200 kg (4800 lb).

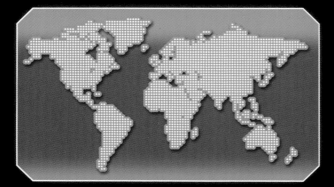

DISTRIBUTION All the temperate and tropical
oceans of the world.

HABITAT Most of the time the sunfish live in
open water at depths of about 200 m (650 ft),
but they are seen at the surface too.

DIET Sunfish love jellyfish and Portuguese
man-of-wars, but they also eat plankton,
crustaceans, small fish and squid.

DESCRIPTION This silver-grey fish has a huge
head and a body that is twice as deep as it is
long. There are long dorsal (upper) and ventral
(lower) fins. The fish's skin is rough like sandpaper. It lacks scales and is covered in slippery
mucus. Inside the fish's small, round mouth are
teeth that are fused together to form a plate
used to grind food.

RELATIVES The ocean sunfish belongs to
the family Molidae, which includes three
species of sunfish: the common sunfish
described here, the sharp-tailed
sunfish and the slender sunfish.

BEHAVIOUR The females lay more
eggs than any other species of fish.
More than 300 million eggs have
been found inside a female. The eggs
are tiny and, after spawning, they
float at the water surface. Here they
hatch into tiny larvae covered in spines.
The spines disappear as the larvae grow
and begin to look more like the adult.

UGE
m
FT)

CLAVUS

The ocean sunfish does not have
a proper tail fin. Instead it has a
rounded end to the tail, called
a clavus. This gives the fish its
very unusual profile.

SIZE
1.6–3 M
5–10 FT

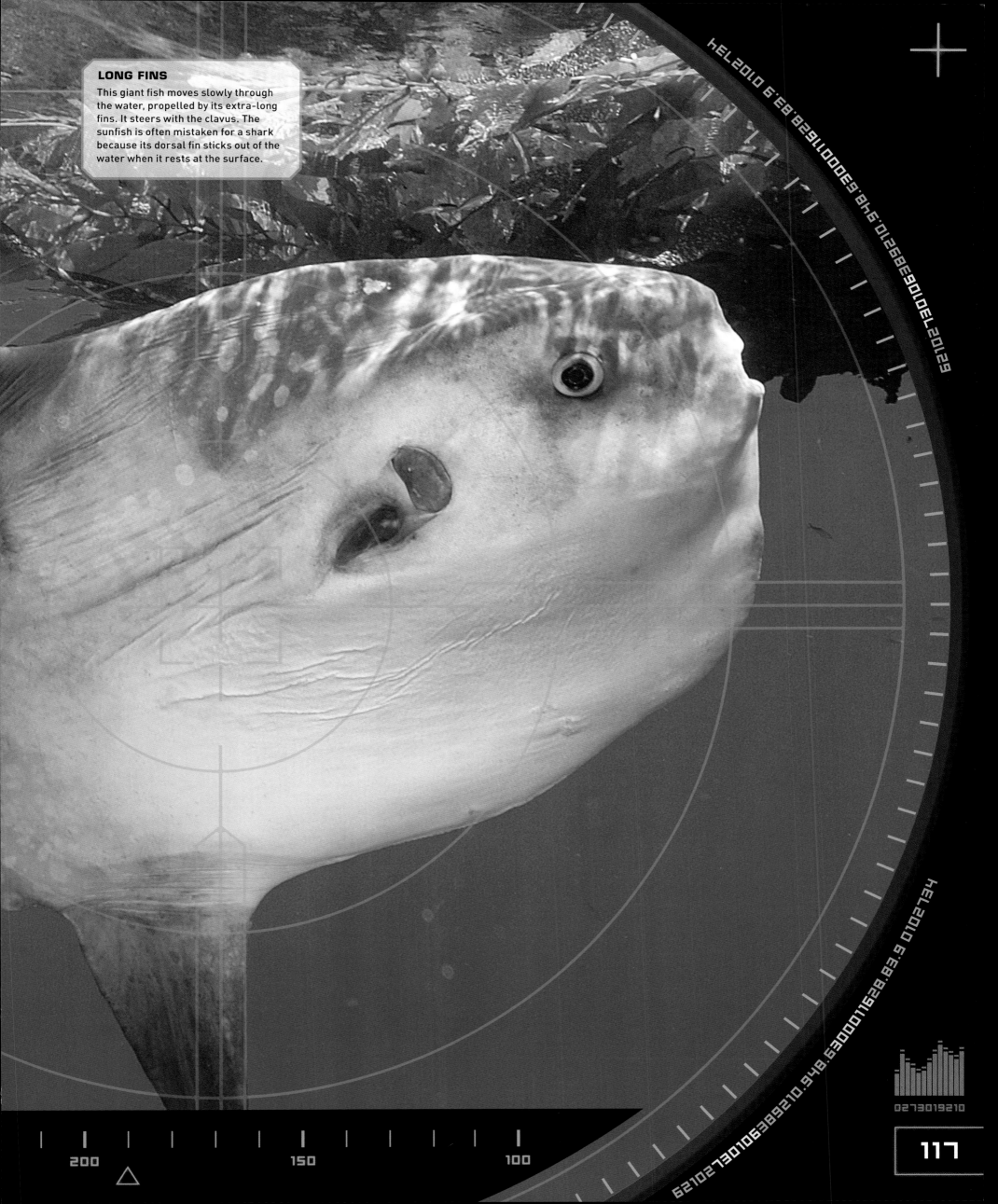

LONG FINS

This giant fish moves slowly through the water, propelled by its extra-long fins. It steers with the clavus. The sunfish is often mistaken for a shark because its dorsal fin sticks out of the water when it rests at the surface.

200   150   100

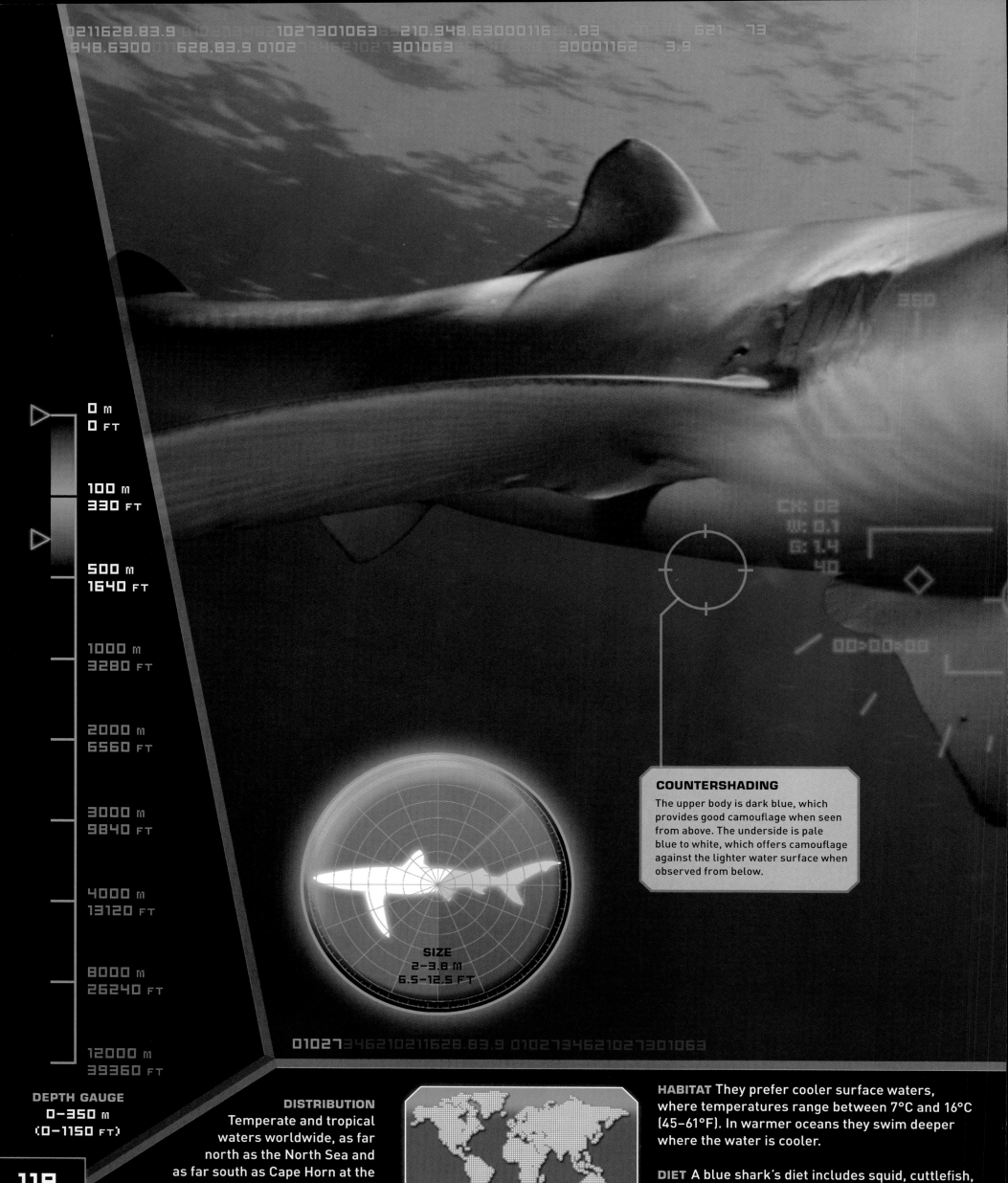

0 m
0 FT

100 m
330 FT

500 m
1640 FT

1000 m
3280 FT

2000 m
6560 FT

3000 m
9840 FT

4000 m
13120 FT

8000 m
26240 FT

12000 m
39360 FT

DEPTH GAUGE
0–350 m
(0–1150 FT)

118

CH: 02
W: 0.1
G: 1.4
40

00>00>00

COUNTERSHADING

The upper body is dark blue, which provides good camouflage when seen from above. The underside is pale blue to white, which offers camouflage against the lighter water surface when observed from below.

SIZE
2–3.8 M
6.5–12.5 FT

0102734621021162 8.83.9 01027 3462102 7301063

DISTRIBUTION

Temperate and tropical waters worldwide, as far north as the North Sea and as far south as Cape Horn at the southern tip of South America.

HABITAT They prefer cooler surface waters, where temperatures range between 7°C and 16°C (45–61°F). In warmer oceans they swim deeper where the water is cooler.

DIET A blue shark's diet includes squid, cuttlefish, octopus, crustaceans, fish and small sharks.

JAGGED TEETH

The blue shark has sharp, jagged-edged teeth, ideal for tearing into flesh. With its great speed, this is a vicious predator, often feeding on schools of fish such as herring, anchovies and sardines. The blue shark has also been known to attack humans, but it rarely ventures into coastal waters.

BLUE SHARK

LATIN NAME *Prionace glauca*

The blue shark is one of the most common large sharks. It is easily identified by its vivid blue body and long pectoral (side) fins. This is one of the fastest-swimming sharks. Its normal cruising speed is about 35 kph (22 mph), but it can reach speeds of up to 100 kph (60 mph) in short bursts.

BEHAVIOUR
The female is viviparous, which means that she gives birth to live young. The eggs remain in her body, where they are fed for up to 12 months. She gives birth to up to 130 pups, each about 50 cm (20 in) in length.

RANGEFINDER 4 m (13 FT)

DESCRIPTION The blue shark has a long, slim body that is dark blue on top and pale blue to white below. Its pectoral fins are long. It has a pointed snout and relatively large eyes. Its jagged teeth are triangular-shaped. The blue shark's average weight is about 200 kg (440 lb).

RELATIVES The blue shark is a member of the Carcharhinidae family, known as the requiem sharks (named after the French word for shark, *requin*). The family includes the reef, bull and dusky sharks. All these sharks give birth to live young. The largest member of the family is the tiger shark, at up to 7.5 m (25 ft) long.

THE DEEP

You have descended to the dark zone. The water is cold and the pressure is so great that if you left the submersible you would be crushed instantly.

No sunlight reaches this far beneath the ocean's surface. The only way to see anything is to use searchlights to find some of the strange animals that live in this dark world. They include the viperfish with its mouthful of fang-like teeth, the gulper eel that is all mouth and stomach, and the anglerfish complete with its own fishing rod.

Some animals, such as the sperm whale, have dived down from the surface. Sperm whales feed on giant squid that are found in the deep, but the whales cannot stay long as they have to return to the surface to breathe. Other animals, such as deep sea crabs, never leave the pitch-dark seabed. As you journey deeper and deeper, remember that more people have travelled in space than have travelled to the bottom of the ocean. Only one manned expedition has visited the deepest spot on Earth, in the Mariana Trench in the northwestern Pacific Ocean.

DEEP MYSTERIES

Little is known about the behaviour of deep sea animals such as this cirrate octopus, but each year our knowledge improves as researchers dive to the depths to film the animals in their environment.

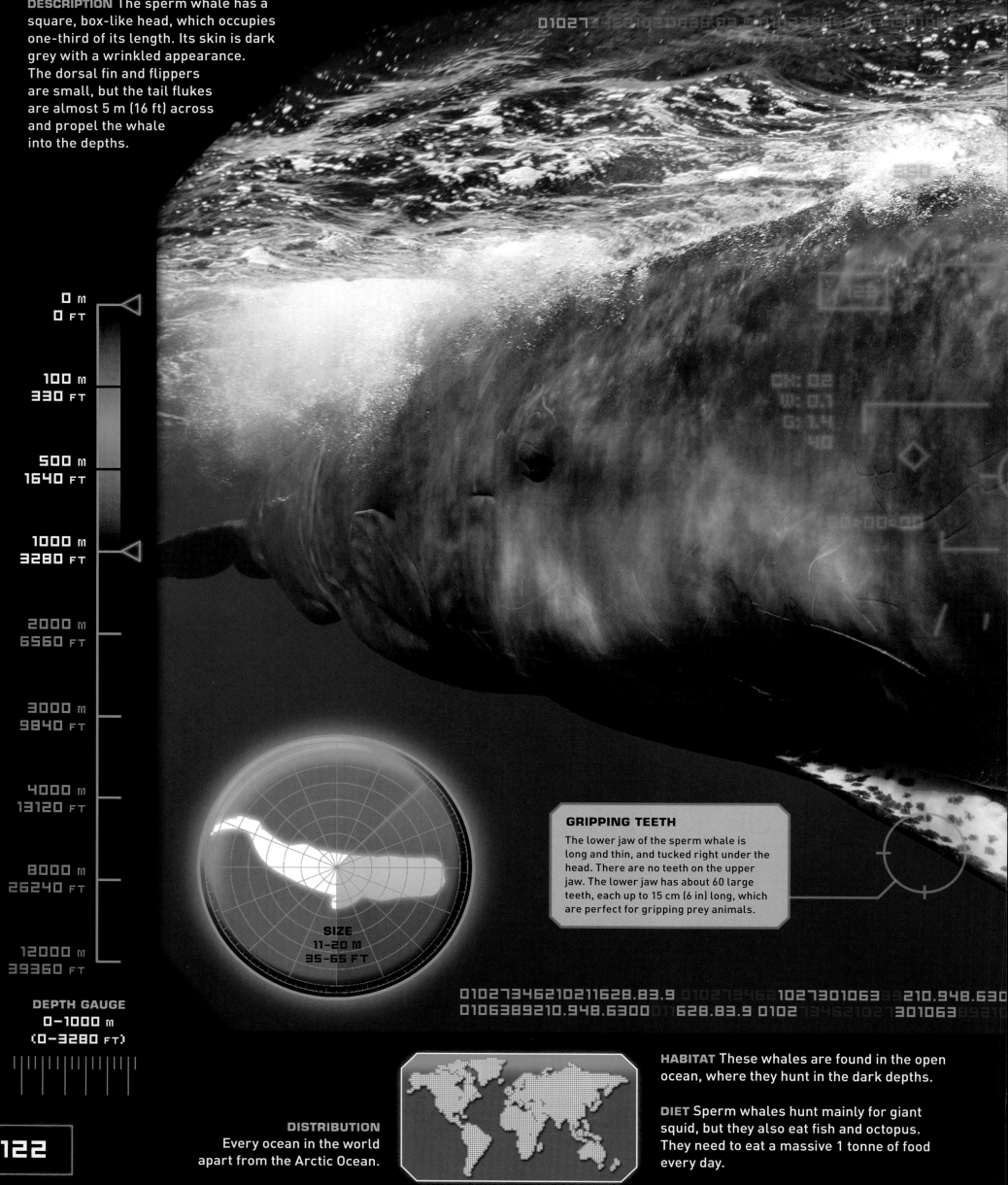

DESCRIPTION The sperm whale has a square, box-like head, which occupies one-third of its length. Its skin is dark grey with a wrinkled appearance. The dorsal fin and flippers are small, but the tail flukes are almost 5 m (16 ft) across and propel the whale into the depths.

0 m
0 FT

100 m
330 FT

500 m
1640 FT

1000 m
3280 FT

2000 m
6560 FT

3000 m
9840 FT

4000 m
13120 FT

8000 m
26240 FT

12000 m
39360 FT

DEPTH GAUGE
0-1000 m
(0-3280 FT)

SIZE
11-20 M
35-65 FT

GRIPPING TEETH

The lower jaw of the sperm whale is long and thin, and tucked right under the head. There are no teeth on the upper jaw. The lower jaw has about 60 large teeth, each up to 15 cm (6 in) long, which are perfect for gripping prey animals.

DISTRIBUTION
Every ocean in the world apart from the Arctic Ocean.

HABITAT These whales are found in the open ocean, where they hunt in the dark depths.

DIET Sperm whales hunt mainly for giant squid, but they also eat fish and octopus. They need to eat a massive 1 tonne of food every day.

SPERM WHALE

LATIN NAME *Physeter macrocephalus*

The square-headed sperm whale is the largest of the toothed whales. It is named after the spermaceti oil that is found in its head and for which it was once hunted. The sperm whale is the deepest-diving whale, reaching depths of 1000 m (3300 ft) or more, and staying under water for up to 90 minutes. When diving in the dark depths, it relies on echolocation to navigate and to find prey.

BATTLE SCARS

The head bears scars, including circular scars made by the giant squid that these whales hunt. In 1965, a sperm whale was found strangled, with the tentacles of a squid wrapped around its head, and the squid's head in its stomach.

BEHAVIOUR The females and their young live in family groups, called pods, of between 10 and 20 individuals. The females share nursery duties while other females go hunting. They stay in breeding grounds close to the Equator. The males spend part of the year with the females and the rest with other males in feeding grounds.

RELATIVES The sperm whale is a toothed whale, like the dolphins, orcas and belugas. Toothed whales have cone-shaped teeth and a single blowhole on the top of their head. The other group of whales, the baleen whales, have baleen plates for filter feeding and two blowholes.

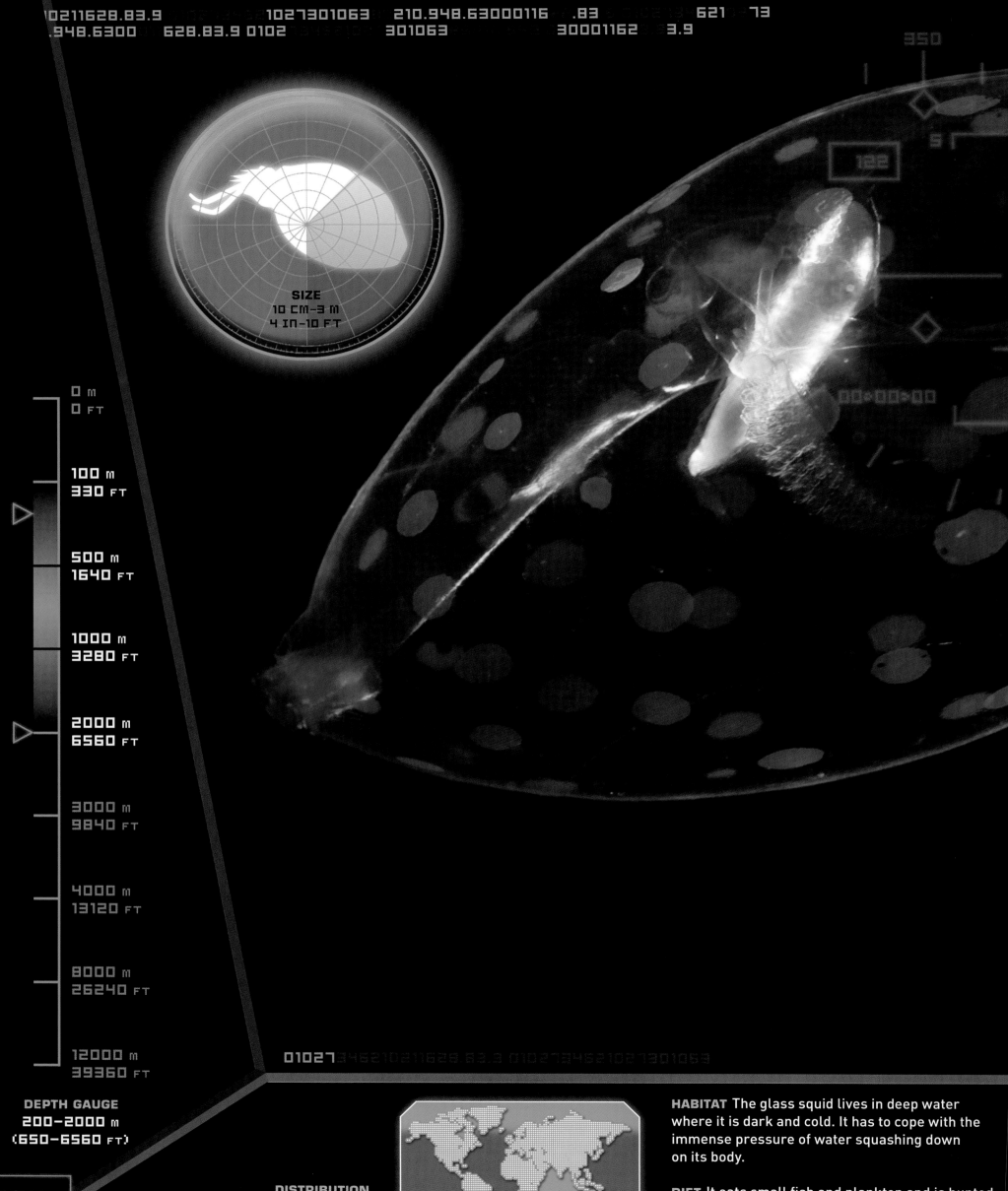

350

122

00:00:00

SIZE
10 CM–3 M
4 IN–10 FT

0 m
0 FT

100 m
330 FT

500 m
1640 FT

1000 m
3280 FT

2000 m
6560 FT

3000 m
9840 FT

4000 m
13120 FT

8000 m
26240 FT

12000 m
39360 FT

01027 346210211628.82.3 01027346210273010653

DEPTH GAUGE
200–2000 m
(650–6560 FT)

DISTRIBUTION
The Pacific and Atlantic Oceans.

HABITAT The glass squid lives in deep water where it is dark and cold. It has to cope with the immense pressure of water squashing down on its body.

DIET It eats small fish and plankton and is hunted by whales and sharks.

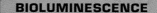

BIOLUMINESCENCE

The glass squid can make its own light by a process called bioluminescence. The light is produced as a result of chemical reactions that take place inside the light-producing organ.

GRIPPING TENTACLES

The glass squid has eight tentacles, two of which are longer than the others. The suckers on the tentacles are used to grip prey. The tentacles then pass the food into the squid's mouth.

DEEP SEA GLASS SQUID

LATIN NAME *Teuthowenia pellucida*

This species of glass squid was found on an expedition to an underwater mountain ridge deep in the Atlantic Ocean. The transparent covering to its body makes it very difficult to spot in the water. When threatened, the squid rolls up into a ball with its tentacles sticking out, a bit like a hedgehog with its spines sticking out. It has a light-producing organ just under each eye.

BEHAVIOUR

The adults live in deep water, but many of the larvae are found in the surface waters, where they live among plankton. As they get older and larger, they descend into deeper water. The larvae look nothing like the adults, but they change shape as they grow.

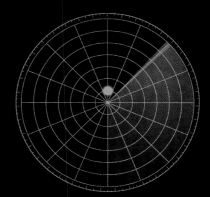

RANGEFINDER 1 m (3 FT)

DESCRIPTION This glass squid has a see-through body that is rounded in shape and covered with brown dots. There are eight striped tentacles, each of which has two rows of suckers. The light-making organs under the squid's eyes are used to cancel out the effects of the animal's own shadow.

RELATIVES Glass squid belong to the family Cranchiidae, which includes about 60 species of glass squid that are found in both surface and deep waters. Most glass squid are transparent: the only organ visible inside their body is the cigar-shaped digestive gland. The largest glass squid is the colossal squid.

0273019210

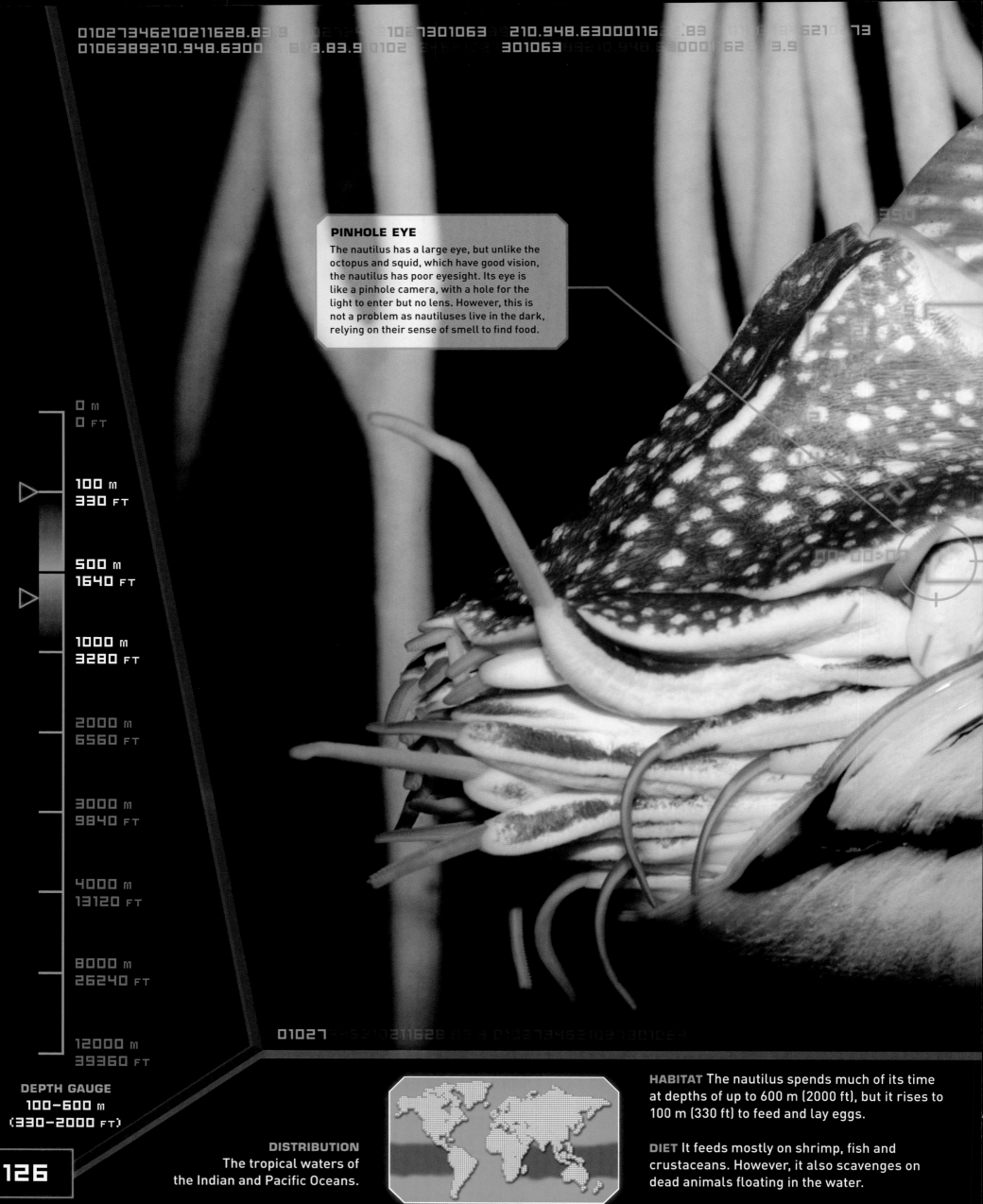

PINHOLE EYE

The nautilus has a large eye, but unlike the octopus and squid, which have good vision, the nautilus has poor eyesight. Its eye is like a pinhole camera, with a hole for the light to enter but no lens. However, this is not a problem as nautiluses live in the dark, relying on their sense of smell to find food.

0 m
0 FT

100 m
330 FT

500 m
1640 FT

1000 m
3280 FT

2000 m
6560 FT

3000 m
9840 FT

4000 m
13120 FT

8000 m
26240 FT

12000 m
39360 FT

DEPTH GAUGE
100–600 m
(330–2000 FT)

DISTRIBUTION
The tropical waters of
the Indian and Pacific Oceans.

HABITAT The nautilus spends much of its time at depths of up to 600 m (2000 ft), but it rises to 100 m (330 ft) to feed and lay eggs.

DIET It feeds mostly on shrimp, fish and crustaceans. However, it also scavenges on dead animals floating in the water.

CHAMBERED NAUTILUS

LATIN NAME *Nautilus pompilius*

The nautilus has a beautiful spiral shell, which protects the soft body underneath it. When threatened, it can withdraw its body completely into the shell. There are separate chambers inside the shell. The baby nautilus has a four-chambered shell, and adds new chambers as it grows until there are about 30 of them. Its body is found in the outermost chamber. The nautilus has a long life of about 20 years.

CAMOUFLAGE

The colours of its shell help to disguise the nautilus in the water. The shell is darker on the upper surface and almost white underneath. This makes it difficult to spot from both above and below.

BEHAVIOUR

The nautilus uses jet propulsion to get around, by drawing water in and out of one of its chambers. It rises and sinks in the water by changing the amount of gas inside its shell. It adds gas to rise up in the water and removes gas to sink.

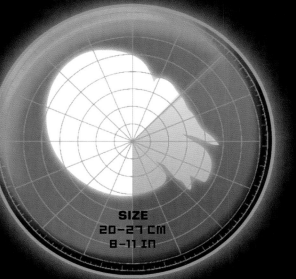

SIZE
20–27 CM
8–11 IN

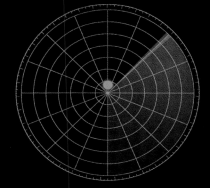

RANGEFINDER
10 CM (4 IN)

DESCRIPTION The nautilus's spiral shell is creamy white with red stripes. This creature has 90 tentacles. Each tentacle has grooves and ridges – rather than suckers – with which to grip prey. A tough, shield-like plate lies over the nautilus's head to offer protection. The parrot-like beak crushes prey.

RELATIVES The nautilus is a type of mollusc known as a cephalopod, with a large head and tentacles, like the squid and octopus. There are seven species of nautilus. They are often called living fossils because they have lived in the oceans for more than 400 million years with little change in their appearance.

VIPERFISH

LATIN NAME *Chauliodus sloani*

With a mouth full of vicious-looking teeth, it is not surprising that the viperfish is one of the fiercest predators of the deep. It swims at full speed towards its prey, stabbing the animal with its teeth. The force from the collision is so great that the first vertebra in the viperfish's backbone acts as a shock absorber to protect the body from damage.

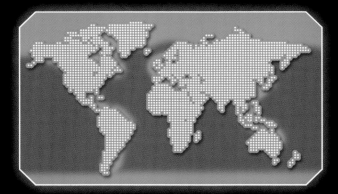

| 0 m
0 FT |
| 100 m
330 FT |
| 500 m
1640 FT |
| 1000 m
3280 FT |
| 2000 m
6560 FT |
| 3000 m
9840 FT |
| 4000 m
13120 FT |
| 8000 m
26240 FT |
| 12000 m
39360 FT |

DEPTH GAUGE
500-2800 m
(1600-9000 FT)

DISTRIBUTION Temperate and tropical oceans across the world.

HABITAT During the day the viperfish stays in deep water about 1500 m (5000 ft) down. When darkness falls, it moves upwards to about 500 m (1600 ft) to find food.

DIET Viperfish eat a variety of crustaceans and fish found in the deep. They can survive for days without feeding.

DESCRIPTION This is a surprisingly small fish, with a long body and an extra-large head. The viperfish is silvery blue to black in colour. There are small light-producing organs under the eyes and along the sides of the body which help to disguise the fish's shape in the water.

RELATIVES The viperfish belong to the family of barbeled dragonfish. There are more than 200 species of dragonfish and they all have long, slender bodies and large heads. These fearsome predators all have a mouth full of fangs.

BEHAVIOUR There is a single, long dorsal spine with a light-producing organ at the end. The fish wiggles the spine and flashes the light on and off to lure prey.

010273462102116̄28.83.8 010273462102730̄

HUGE HEAD
There is little prey swimming in the deep ocean, so this fish has to be able to catch and swallow almost any prey that comes close. The large mouth and stomach allow it to grab animals that are the same size as itself.

SIZE
20-30 CM
8-12 IN

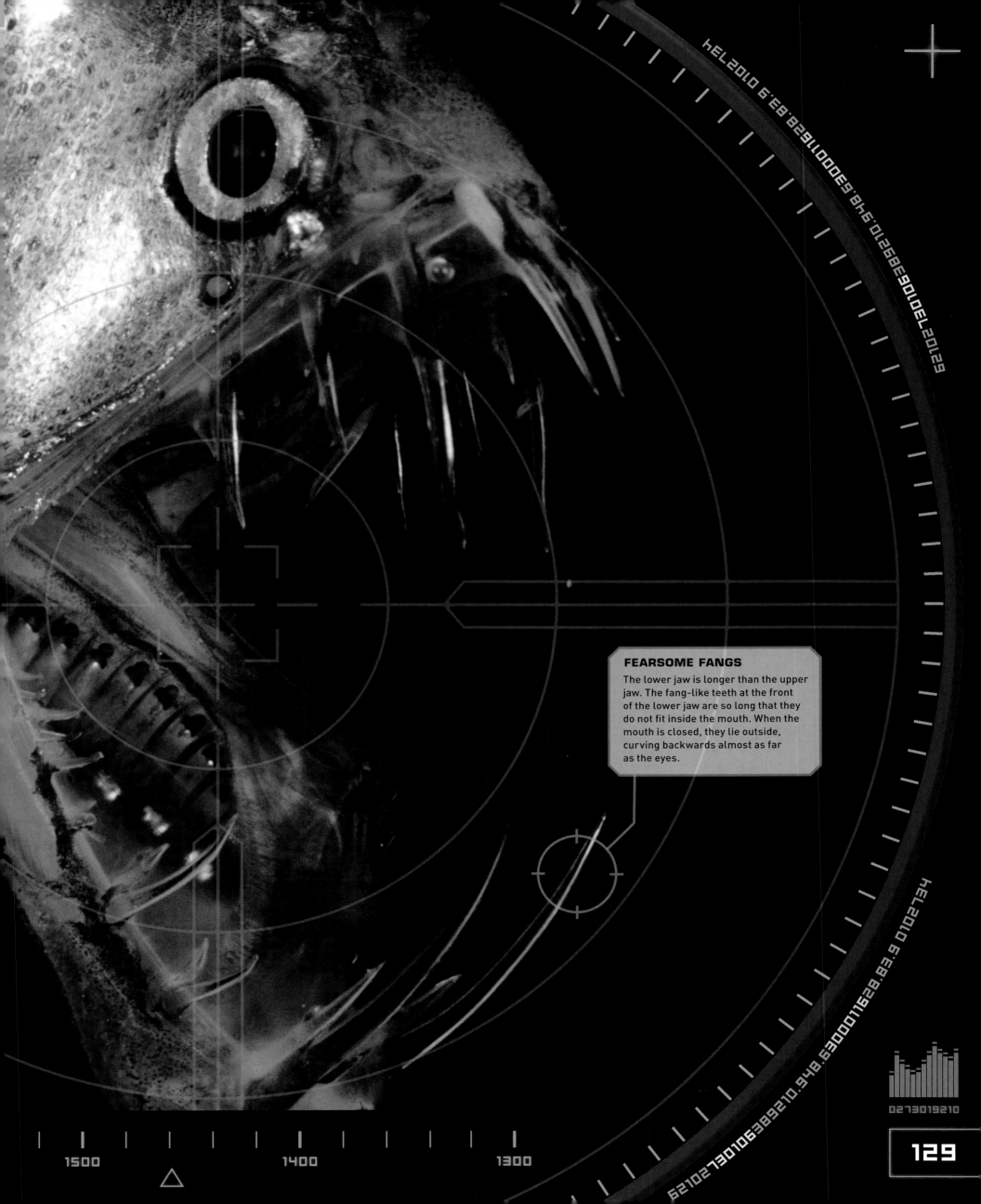

FEARSOME FANGS

The lower jaw is longer than the upper jaw. The fang-like teeth at the front of the lower jaw are so long that they do not fit inside the mouth. When the mouth is closed, they lie outside, curving backwards almost as far as the eyes.

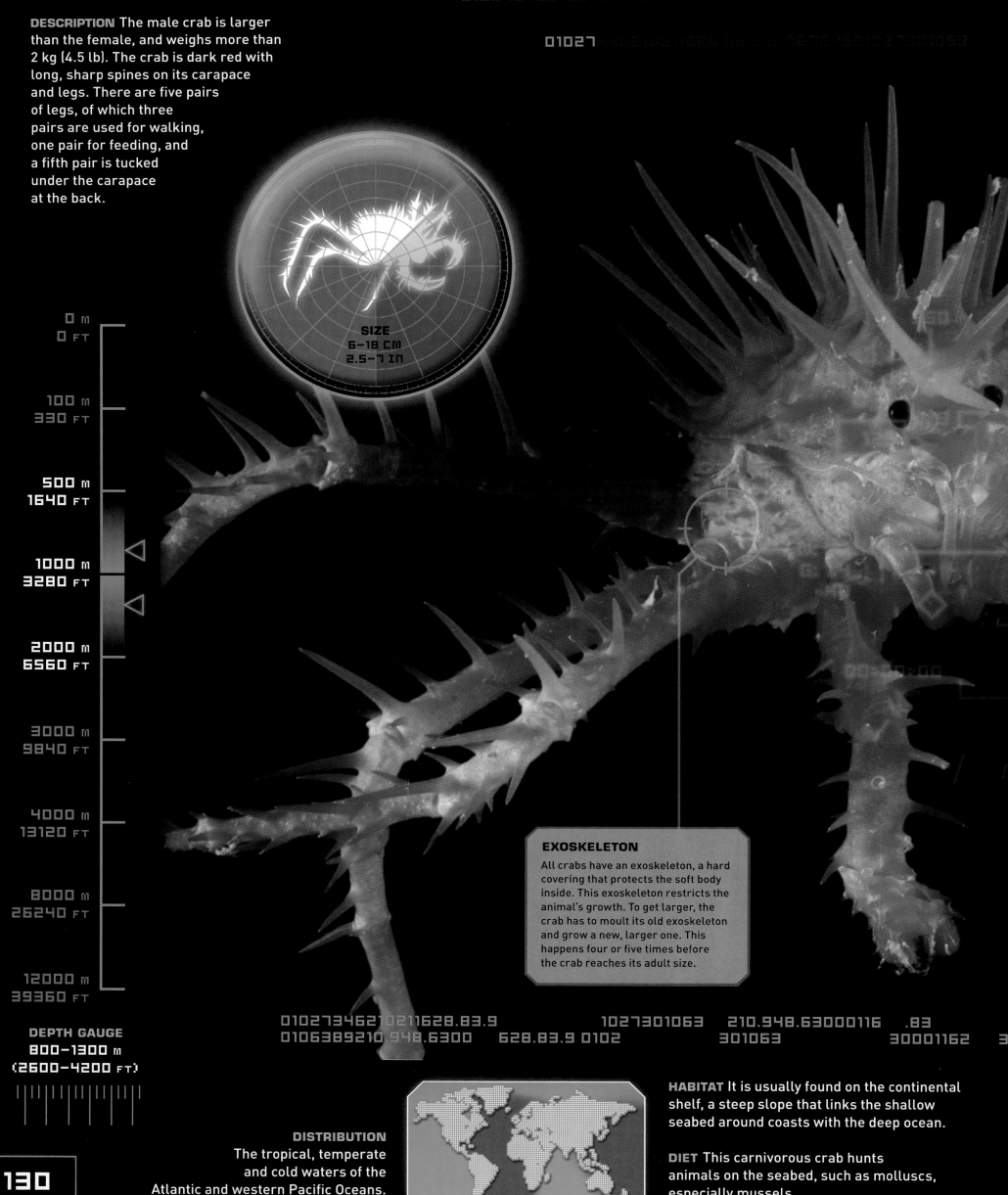

DESCRIPTION The male crab is larger than the female, and weighs more than 2 kg (4.5 lb). The crab is dark red with long, sharp spines on its carapace and legs. There are five pairs of legs, of which three pairs are used for walking, one pair for feeding, and a fifth pair is tucked under the carapace at the back.

SIZE
6–18 CM
2.5–7 IN

0 m
0 FT

100 m
330 FT

500 m
1640 FT

1000 m
3280 FT

2000 m
6560 FT

3000 m
9840 FT

4000 m
13120 FT

8000 m
26240 FT

12000 m
39360 FT

EXOSKELETON

All crabs have an exoskeleton, a hard covering that protects the soft body inside. This exoskeleton restricts the animal's growth. To get larger, the crab has to moult its old exoskeleton and grow a new, larger one. This happens four or five times before the crab reaches its adult size.

010273462102116 28.83.9   1027301063   210.948.63000116   .83
0106389210.948.6300   628.83.9 0102   301063   30001162   3.

DEPTH GAUGE
800–1300 m
(2600–4200 FT)

DISTRIBUTION
The tropical, temperate and cold waters of the Atlantic and western Pacific Oceans.

HABITAT It is usually found on the continental shelf, a steep slope that links the shallow seabed around coasts with the deep ocean.

DIET This carnivorous crab hunts animals on the seabed, such as molluscs, especially mussels.

DEEP SEA CRAB

LATIN NAME *Neolithodes* sp.

This red crab is covered in sharp spines, which is why it is often called the porcupine crab. Crabs have ten legs: four pairs of walking legs and one pair with larger claws for feeding. This crab looks as if it has just three pairs of walking legs, but there is a pair tucked away out of sight under the carapace (shell). The crab uses this small pair of legs to clean its gills. Little is known about these deep sea crabs, although they are often caught in the nets used to catch turbot, a bottom-dwelling flat fish.

LONG LEGS

The crab uses its three pairs of long legs to walk over the seabed in search of invertebrate animals. The sharp spines provide protection against predators.

BEHAVIOUR The front pair of legs has larger claws that are used for catching and manipulating food. The right claw is larger than the left one and is used for crushing food, such as the thick shells of mussels. The smaller claw is used to handle food.

SPECIES The deep sea crab is a member of the king crab, or stone crab, family, of which there are 79 species. There are about 10 species of crab in the genus *Neolithodes*. They are all deep sea species, but they are found in different parts of the world.

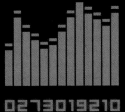

RANGEFINDER 5 cm (2 in

131

28.83.9 010273462102730106389210.948.63000116 28.83.9 010273 46621027 3
i300 628.83.9 0102 134621027 30106389210.948. 63000116 28.83.9

ANGLERFISH

LATIN NAME *Melanocetus* sp.

The anglerfish gets its name from the way the female hunts. Three long spines stick up from between its eyes. The longest spine is the 'fishing rod'. At the end of the fishing rod is a fleshy knob that can be wiggled to attract prey. An anglerfish has a wide, gaping mouth that snaps shut as soon as anything touches it. Its long, pointed teeth face backwards, so a fish caught in them only impales itself deeper as it tries to escape.

DISTRIBUTION Found mostly in the Atlantic and Southern Oceans.

HABITAT These are fish that can cope with living in the dark, cold water of the deepest parts of the ocean.

DIET Anglerfish are not fussy eaters and will catch any fish up to twice their size.

DESCRIPTION The female fish varies in colour from red-brown to dark grey. It has a huge head, with a wide mouth and sharp teeth. Its rounded body looks a bit like a basketball.

SPECIES There are about 200 species of anglerfish, and most live in the deep ocean. Some are fished commercially – the flesh is said to taste like that of a lobster. The numbers of anglerfish are falling because they are currently being overfished.

BEHAVIOUR Anglerfish are found scattered through the ocean, so finding a mate could be tricky. To avoid the problem, the female carries around a tiny male with her. A young male swims freely, but once it finds a female, it attaches to her permanently. The male bites through her skin to reach a blood vessel and get food.

0 m
0 FT

100 m
330 FT

500 m
1640 FT

1000 m
3280 FT

2000 m
6560 FT

3000 m
9840 FT

4000 m
13120 FT

8000 m
26240 FT

12000 m
39360 FT

DEPTH GAUGE
1000–4000 m
(3280–13120 FT)

SIZE
20–100 CM
8–40 IN

01027 3462102116 28.83.9

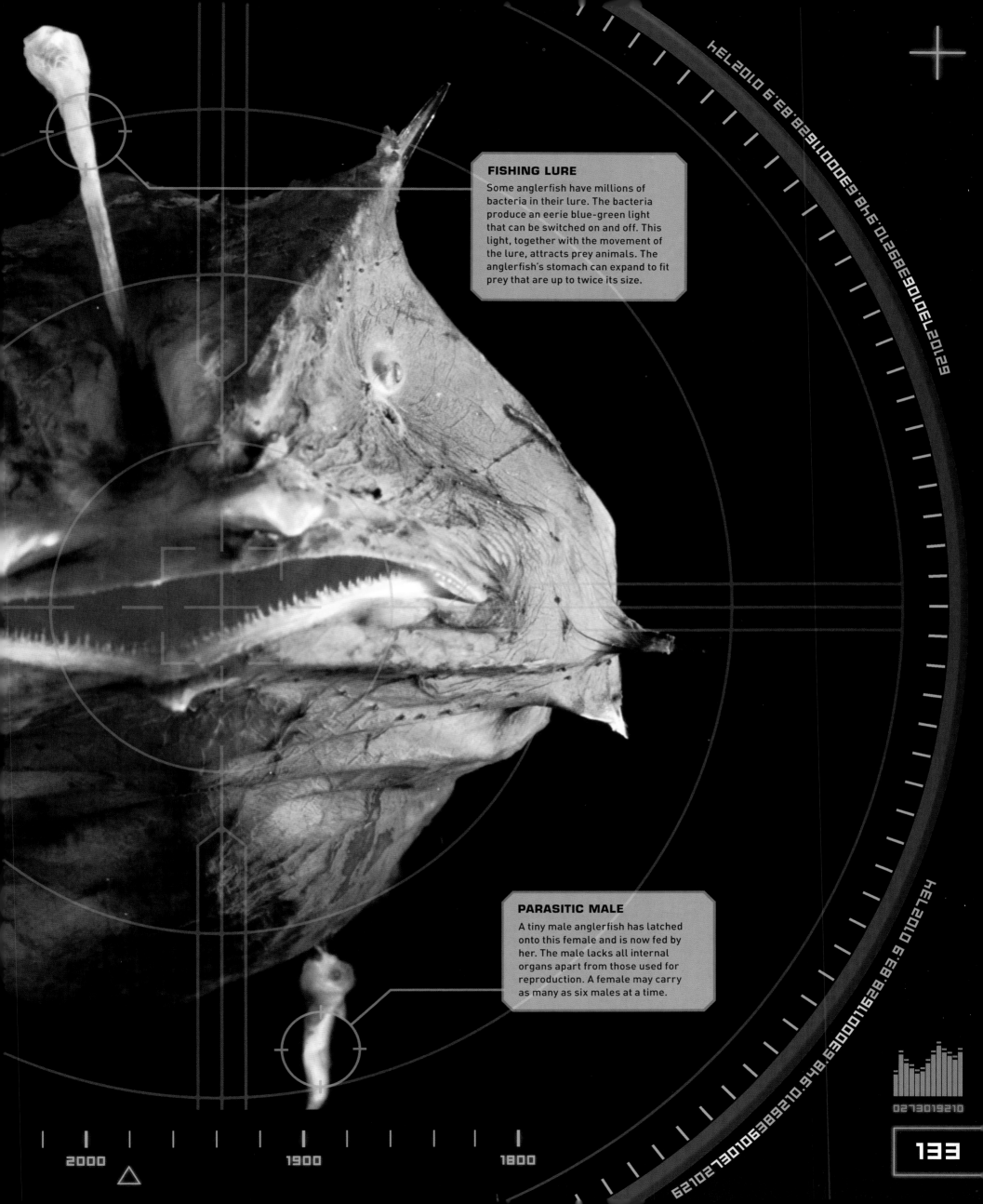

FISHING LURE

Some anglerfish have millions of bacteria in their lure. The bacteria produce an eerie blue-green light that can be switched on and off. This light, together with the movement of the lure, attracts prey animals. The anglerfish's stomach can expand to fit prey that are up to twice its size.

PARASITIC MALE

A tiny male anglerfish has latched onto this female and is now fed by her. The male lacks all internal organs apart from those used for reproduction. A female may carry as many as six males at a time.

DEEP SEA OCTOPUS

LATIN NAME *Sauroteuthis syrtensis*

This deep sea octopus is often called the Dumbo octopus because of the ear-like fins behind its eyes. It is a type of cirrate octopus because it has tiny bristles, called cirri, on its tentacles. The cirri can move backwards and forwards to create small currents in the water that pull tiny animals towards the octopus's arms.

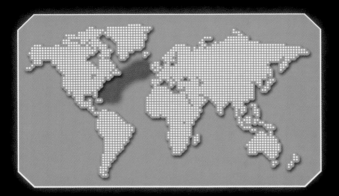

DISTRIBUTION The North Atlantic Ocean, especially off the eastern coast of North America.

HABITAT It is often seen on the continental shelf up to depths of 4000 m (13000 ft), but it is most common at 1500–2500 m (5000–8000 ft).

DIET The octopus feeds on small animals in the water, such as plankton and tiny crustaceans, especially copepods, and it may even scavenge for dead creatures on the seabed.

DESCRIPTION It has eight tentacles with suckers. Males have large and small suckers, but the females just have small ones. There is webbing between the tentacles, which forms an umbrella when the tentacles are extended. There are two flap-like fins on the head.

RELATIVES There are 45 species of cirrate octopus, most of which live at depths of 300 m (1000 ft) or more. A few are found in shallow water in the Arctic. Only one other species can produce bioluminescence like the species pictured here.

BEHAVIOUR Most octopuses use their suckers to catch and grip food, but this deep sea species is different. It has adapted to the darkness of the deep by changing its suckers. The suckers cannot stick to anything, but produce a blue-green light instead.

0 m / 0 FT
100 m / 330 FT
500 m / 1640 FT
1000 m / 3280 FT
2000 m / 6560 FT
3000 m / 9840 FT
4000 m / 13120 FT
8000 m / 26240 FT
12000 m / 39360 FT

DEPTH GAUGE
500–4000 m
(1640–13120 FT)

WEBBED TENTACLES
Skin extends between the tentacles, forming a web that is held out like an umbrella. The light produced by the suckers lures prey animals. They are trapped inside the umbrella and pushed into the beak-like mouth.

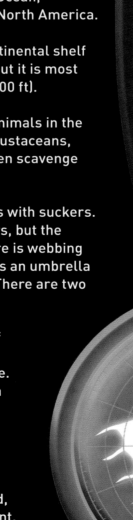

01027

SIZE
25–50 CM
10–20 IN

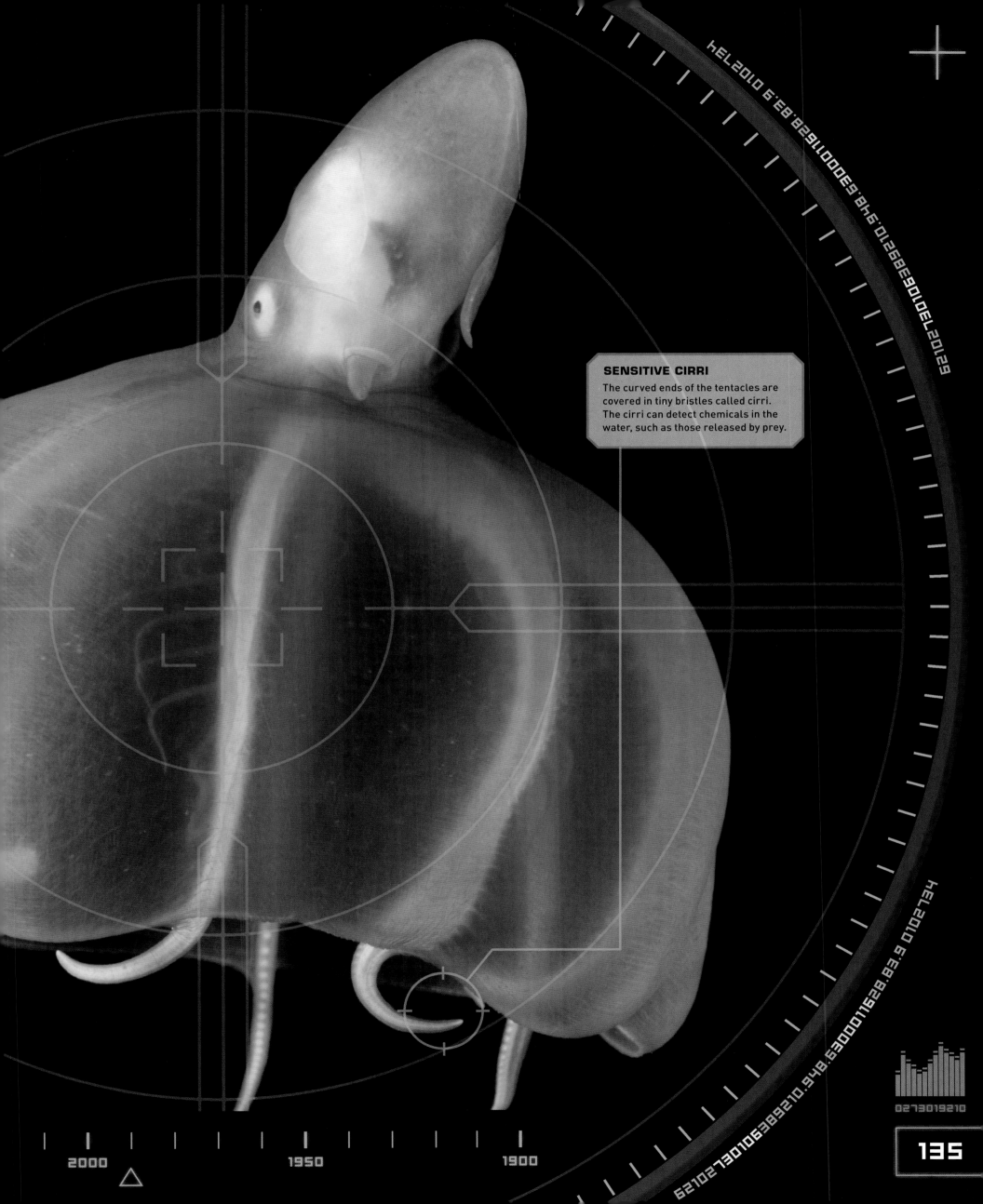

SENSITIVE CIRRI

The curved ends of the tentacles are covered in tiny bristles called cirri. The cirri can detect chemicals in the water, such as those released by prey.

SIZE
80–100 CM
32–40 IN

CATCHING FOOD

When a prey animal comes close, the gulper eel
lunges forwards and opens its mouth. Water
rushes into the huge mouth, sweeping the prey
with it. The mouth shuts as quickly as it opened,
and the water leaves though the gill slits.

350

122

CH:
W:

00:00:00

0 m
0 FT

100 m
330 FT

500 m
1640 FT

1000 m
3280 FT

2000 m
6560 FT

3000 m
9840 FT

4000 m
13120 FT

8000 m
26240 FT

12000 m
39360 FT

01027346210211628.83.9 0102734621027301063

DEPTH GAUGE
500–7700 m
(1640–25000 FT)

136

DISTRIBUTION
Around the world, in both
temperate and tropical waters.

HABITAT The deepest ocean. A gulper eel has
been found as deep as 7700m (25000 ft), but they
live mostly at about 1300 m (4300 ft).

DIET Gulper eels can swallow large prey, but
they feed mostly on small crustaceans, fish,
squid and octopus.

BIO-LIGHT

The gulper eel has a light-producing knob at the end of its tail, probably in order to attract prey. Sometimes the eel waves its tail in front of its mouth. At other times, it chases it round and round in circles.

GULPER EEL

LATIN NAME *Eurypharynx pelecanoides*

This bizarre deep sea fish does not really look like a fish. Its mouth is larger than its body and its jaws are loosely hinged so they can open really wide. It has an extra-long whip-like tail with a light-producing organ at the end. This creature is known as the pelican or umbrella gulper eel because of its pouch-like mouth. The adult fish are found in the deepest parts of the ocean, but the young larvae are found near the surface.

BEHAVIOUR

The females release a smelly chemical into the water to attract the males, which use their amazing sense of smell to find them. The eggs hatch into tiny larvae that have small heads and a flat, transparent body. As the larvae grow, they become more like the adult fish.

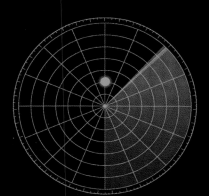

RANGEFINDER 4 m (13 FT)

DESCRIPTION The gulper eel has a huge mouth with extra-long jaws, a small body with a large stomach, and a long thin tail. The fish lacks scales or teeth, and has tiny eyes. The pectoral (side) fins are very small and the tail fin is non-existent. The dorsal (upper) fin extends from behind the head right to the tail.

RELATIVES The gulper eel belongs to the group of swallowers and gulpers. These eel-like fish have a huge mouth and a body that lacks scales, pelvic fins and a swim bladder (a gas-filled organ that helps fish to control their depth). Most swallowers and gulpers live in the deep ocean.

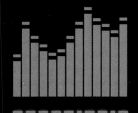

HATCHETFISH

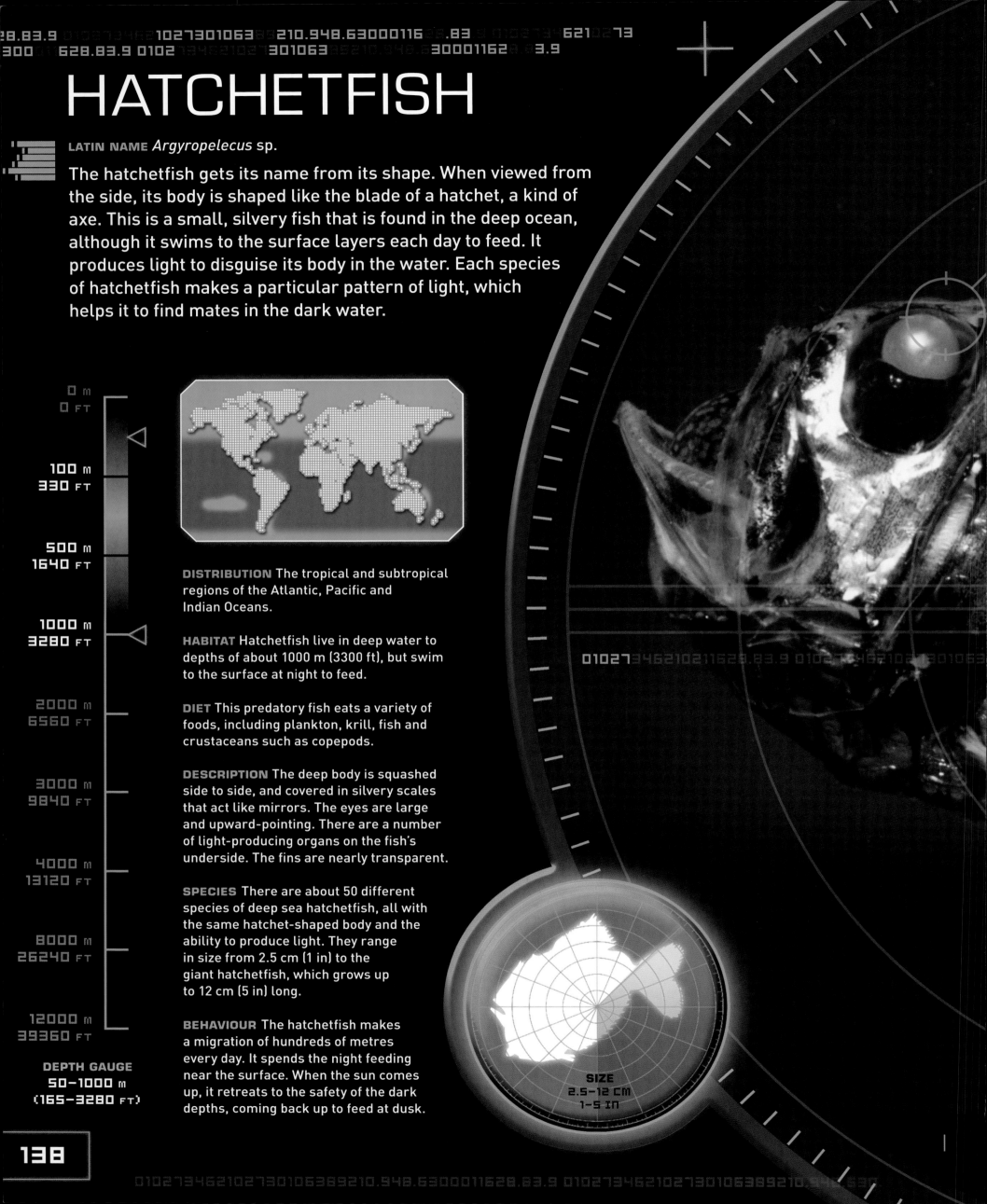

LATIN NAME *Argyropelecus* sp.

The hatchetfish gets its name from its shape. When viewed from the side, its body is shaped like the blade of a hatchet, a kind of axe. This is a small, silvery fish that is found in the deep ocean, although it swims to the surface layers each day to feed. It produces light to disguise its body in the water. Each species of hatchetfish makes a particular pattern of light, which helps it to find mates in the dark water.

DISTRIBUTION The tropical and subtropical regions of the Atlantic, Pacific and Indian Oceans.

HABITAT Hatchetfish live in deep water to depths of about 1000 m (3300 ft), but swim to the surface at night to feed.

DIET This predatory fish eats a variety of foods, including plankton, krill, fish and crustaceans such as copepods.

DESCRIPTION The deep body is squashed side to side, and covered in silvery scales that act like mirrors. The eyes are large and upward-pointing. There are a number of light-producing organs on the fish's underside. The fins are nearly transparent.

SPECIES There are about 50 different species of deep sea hatchetfish, all with the same hatchet-shaped body and the ability to produce light. They range in size from 2.5 cm (1 in) to the giant hatchetfish, which grows up to 12 cm (5 in) long.

BEHAVIOUR The hatchetfish makes a migration of hundreds of metres every day. It spends the night feeding near the surface. When the sun comes up, it retreats to the safety of the dark depths, coming back up to feed at dusk.

0 m
0 FT

100 m
330 FT

500 m
1640 FT

1000 m
3280 FT

2000 m
6560 FT

3000 m
9840 FT

4000 m
13120 FT

8000 m
26240 FT

12000 m
39360 FT

DEPTH GAUGE
50-1000 m
(165-3280 FT)

SIZE
2.5-12 CM
1-5 IN

01027346210211628.83.9 0102734621021301063

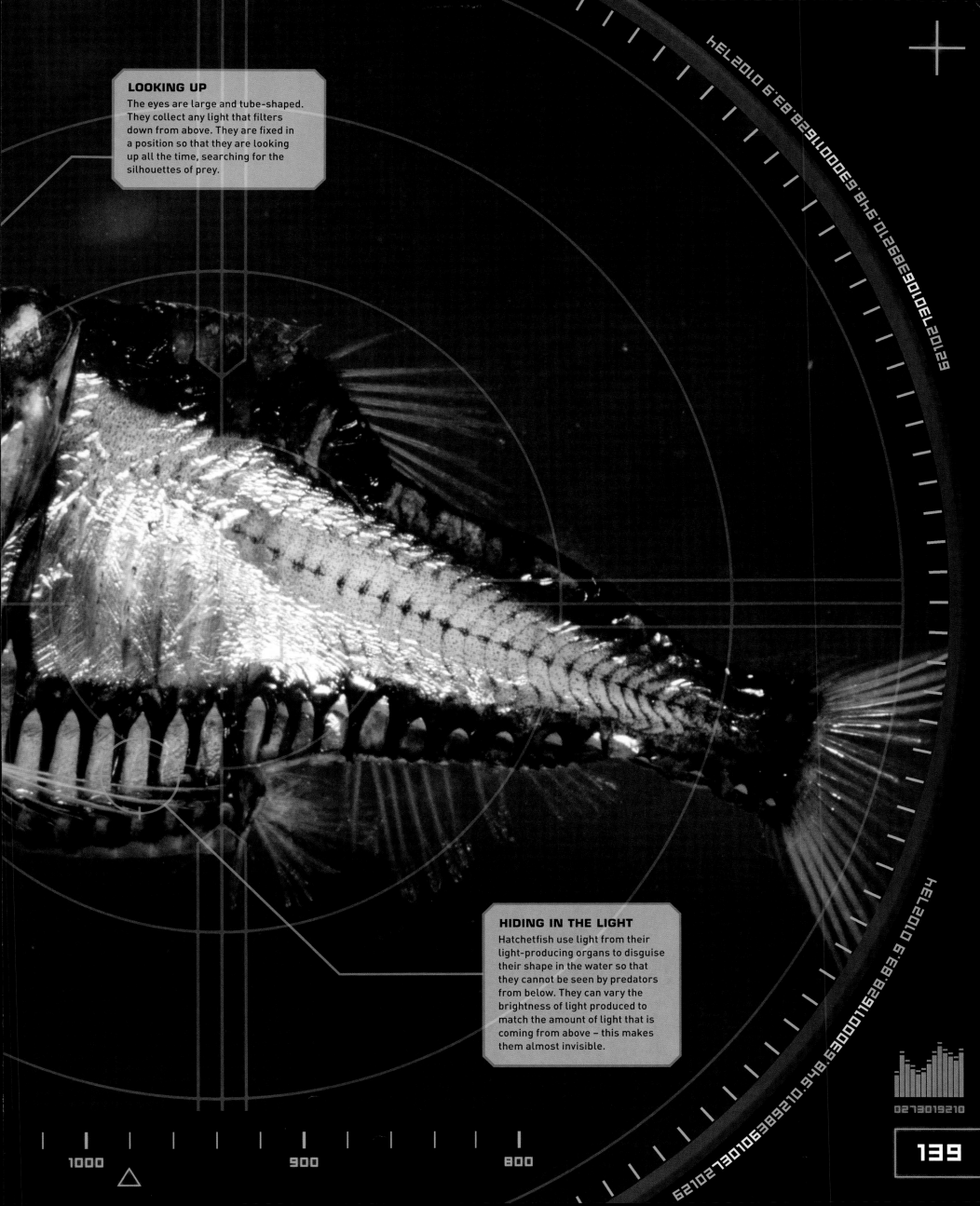

LOOKING UP

The eyes are large and tube-shaped. They collect any light that filters down from above. They are fixed in a position so that they are looking up all the time, searching for the silhouettes of prey.

HIDING IN THE LIGHT

Hatchetfish use light from their light-producing organs to disguise their shape in the water so that they cannot be seen by predators from below. They can vary the brightness of light produced to match the amount of light that is coming from above – this makes them almost invisible.

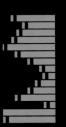

GLOSSARY

ABDOMEN The third or hind region of the body, containing the digestive and reproductive organs.

ALGAE Simple plants such as seaweeds.

AMBUSH Sudden attack.

ANTENNAE The sensory feelers on the head of an insect or crustacean.

AQUATIC Living or growing in water.

BALEEN The large plates in the mouth of some whales, used to sieve food from the water.

BIOLUMINESCENCE Production of light by a living organism.

BIVALVE A type of mollusc with two shells hinged together.

BLUBBER Layer of fat under the skin that traps heat.

BRYOZOAN An invertebrate animal that forms colonies of polyps. Bryozoans build stony skeletons that are a little similar to coral.

CAMOUFLAGE Colouring that disguises an animal's shape so that it blends into the background.

CARAPACE Hard, shell-like covering over the dorsal surface of a crab, turtle or similar animal.

CARNIVORE Animal that eats meat.

CARTILAGINOUS Having a skeleton that is made from bendier cartilage rather than bone.

CEPHALOPOD A type of mollusc with sucker-bearing tentacles.

CETACEAN A large marine mammal such as a whale or a dolphin.

COLONY A group of animals living together.

CONTINENTAL SHELF The edge of a continent that extends out under shallow water.

CORAL Marine animal, related to the sea anemone, that lives in colonies and is an important reef builder.

CRUSTACEAN An invertebrate animal with a hard exoskeleton, such as a crab or a prawn.

DIGESTIVE Something that causes digestion, or the breakdown of food.

DISTRIBUTION The geographical area in which an animal is found.

DORSAL The upper surface of an animal.

ECHINODERM A spiny-skinned invertebrate animal, such as a seastar or sea urchin.

ECHOLOCATION The use of sound to navigate or search for food.

ENDANGERED At risk of becoming extinct, or dying out completely.

EXOSKELETON The hard, protective covering around the body of many invertebrate animals, such as insects.

FAMILY When scientists classify living things into groups, this is a group of related living things ranking below an order and above a genus.

FISH An aquatic vertebrate, covered in scales and with fins for swimming.

FLIPPER The limb of an aquatic animal that is flattened like a paddle for swimming.

GENUS A group of living things belonging to the same family and consisting of a number of species.

GILL RAKERS Comb-like structures inside the gills used to filter food from the water.

GILLS The structures that an aquatic animal, such as a fish, uses to obtain oxygen from the water.

GILL SLITS Holes through which water passes out of the body of a fish, such as a shark or ray.

HABITAT The home of an animal.

HERBIVORE An animal that eats plants.

HERMAPHRODITE With both male and female sex organs.

INSULATION Material that traps heat inside an animal's body, such as fur, feathers or blubber.

INVERTEBRATE An animal without a backbone, such as an insect or mollusc.

IRIDESCENT Colours that vary in appearance when viewed from different angles.

JUVENILE A young animal.

KRILL Shrimp-like animals that are found in large numbers in water, living together in groups. The main food of baleen whales.

LAGOON A large area of saltwater cut off from the ocean by a sand bar or coral reef.

LARVA A young animal, such as a young fish or insect, that is undergoing physical change before it takes on its adult form.

LATERAL FINS Fins positioned at the sides of the body.

LURE Bait, structure used by fish to attract their prey.

MAMMAL An animal that breathes air into lungs, feeds its young with milk and grows some hair.

MARINE Living in saltwater.

010273462102116E8.83.9 010273462102730106389210.948.6300011628.83.9
010273462102730106389210.948.6300011628.83.9 010273462102730106389210

MATURE Adult.

MIGRATION A journey between one habitat and another, made at specific times.

MOLLUSC An animal with a shell and a muscular foot for moving.

MOULT The seasonal loss of hair or feathers, which regrow.

ORDER When scientists classify living things into groups, this is a group of living things that ranks below a class (for example, the mammals) and above a family.

PARASITE An animal that lives in or on another animal, causing that animal harm.

PECTORAL FINS The paired fins of a fish, found behind the head.

PELVIC FINS Paired fins of a fish, found behind the pectoral fins and in front of the tail.

PERISCOPE A viewing device consisting of a tube containing mirrors, used to see something that is not in a direct line of sight.

PLANKTON Microscopic plants and animals that float in the surface layer of the oceans.

POLAR Relating to places near the North or South Poles.

POLYP Part of the life cycle of a coral or a sea anemone. A small animal with a tube-like body and ring of tentacles around the mouth.

POPULATION The number of animals of the same type living in the same area.

PREDATOR An animal that hunts other animals.

PREY An animal that is hunted by a predator.

PROBOSCIS Sucking mouthpart, found in insects such as flies and butterflies.

PROTECTED Conserved, kept safe.

RANGE The distribution of an animal, the areas in which it is found.

RHOPALIA Sense organ found in jellyfish which is sensitive to light and position in the water.

SCAVENGER An animal that feeds on dead and decaying matter.

SCHOOL A group of fish or other aquatic animals that swims in the same direction.

SEA ICE Ice that is formed when sea water freezes.

SEDENTARY Not moving, remaining in the same place.

SENSORY Able to detect sensations or stimuli such as light, smell, sound and taste.

SHOAL A group of fish that stays together, perhaps for defence or for finding a mate easily.

SKELETON The hard framework of an animal's body that gives it support and protection.

SOLITARY Living alone.

SPAWNING Laying eggs.

SPECIES A particular type of animal or plant with a unique appearance, which cannot breed with other types.

SPONGE Simple animal that filters food from the water.

STREAMLINED Having a smooth shape that slips through the water easily.

SUBARCTIC Relating to places just south of the Arctic Circle.

SUBMERSIBLE An underwater vessel for exploring the ocean, a mini submarine.

SUBSPECIES Group within a species that shows minor differences, but not enough to make it a separate species.

SUBTROPICAL Relating to places to the north and south of the tropics that have warm weather.

TEMPERATE Regions that lie between the tropics and the Poles, that experience seasonal weather changes, with weather that is generally cold and wet in winter and warm and dry in summer.

TENTACLE Long, flexible, finger-like structure found in many animals such as anemones and octopuses.

TERRITORY The area in which an animal lives.

TOXIN A poison.

TRANSPARENT See-through.

TROPICAL Relating to places near the Equator that have hot, humid weather for much of the year.

VENOMOUS Poisonous.

VENTRAL The lower surface or underside of an animal.

VERTEBRATE An animal with a backbone.

VIVIPAROUS Giving birth to live young, rather than laying eggs.

WRASSE A fish in the Labridae family, with thick lips, strong teeth and, usually, bright colouring.

0273019210

948.6300011628.83.9
2734621027301063B9210

INDEX

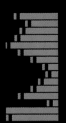

A

abyss 8
ambiguous sea spider 72–73
amphipods 106
anemone hermit crab 60–61
angelfish 46–47
anglerfish 38, 120,
    132–33
Antarctic 76, 77, 81,
    84–87
antennae 32
antifreeze 82, 87
Arctic Ocean 9, 76, 92–95
armour 28
asexual reproduction 26
Atlantic Ocean 8, 82, 125

B

babies 16, 81, 91
baleen whales 78
barracudas 42–43
batfish 38–39
bioluminescence 106, 125,
    128, 133–39
blackfin barracuda 42–43
blood 86, 114
blubber 14, 81, 92,
    95, 103
blue-ringed octopus 52–53
blue shark 99, 118–19
blue whale 98, 108–109
bonnet ray 30–31
breathing 16, 73, 86, 87
brood flap/patch/pouch
    20, 28, 67

C

Californian sea lion 14–15
camouflage 20, 23, 34, 37,
    51, 61, 63, 104, 115,
    118, 127
Caribbean reef shark 44–45
cartilaginous fish 71
cephalopods 22, 127
chambered nautilus 126–27
cirri 134, 135
cleaners 57, 62, 67,
    71, 116
clown fish 40, 56–57
coastal waters 11, 12–39
coasts 11, 97

colour 23, 48, 53
continental shelf 8, 97
coral 40, 41
coral reefs 11, 40–75
crabs 11, 45
    deep sea 120, 130–31
    hermit 60–61
    porcelain 66–67
crocodile icefish 86–87
cross jelly 106–107
crown jellyfish 100–101
cuttlefish 22–23

D

decapods 67
deep ocean 6, 8, 97, 120–39
deep sea crab 120, 130–31
deep sea glass squid
    124–25
deep sea octopus 134–35
dolphins 91, 99, 102–103
dumbo octopus 134–35

E

eagle rays 12, 30–31
echinoderms 24
echolocation 90, 103, 123
eel, moray 62–63
eggs 16, 20, 23, 59, 72,
    82, 116
electroreception 31, 44
emperor penguin 76, 80–81
endangered species 16,
    20, 37, 45
exoskeleton 130
eyes 32, 126, 138

F

fighting 93
filter feeding 108
fins 23, 82, 104, 117
fish
    coastal 18–21, 34–39
    deep-sea 128–29,
        132–33, 136–39
    open ocean 98, 99,
        104–105, 116–17
    polar 82–83, 86–87
    reef 42–47, 50–51, 54–
        57, 62–65, 68–71, 74–75

fishing 45, 103
    overfishing 27, 82
flippers 14, 17, 78, 92
frogfish 104–105
fur 14, 15, 94

G

garibaldi 18–19
giant squid 120, 123
gills 31, 59
glass squid 124–25
global warming 8
great white shark 97,
    114–15
green turtle 16–17
greenthroat parrotfish
    68–69
groupers 64–65
groups 15, 90, 92,
    102, 123
gulper eels 120, 136–37

H

harlequin shrimp 48–49
hatchetfish 138–39
hermaphrodites 59, 68
hermit crabs 60–61
humpback whale 78–79

I J K

ice 76, 77, 88
icefish 86–87
Indian Ocean 9
jellyfish 97, 100–101,
    106–107
jet propulsion 127
jewel anemone 26–27
kelp forests 11, 13
killer whale 90–91
krill 76, 88–89

L

lacy scorpionfish 34–35
larvae 62, 74, 137
leafy seadragon 20–21
leopard seal 76, 84–85
lionfish 40, 50–51
live birth 31, 37, 45,
    85, 111, 115, 119

ACKNOWLEDGEMENTS

Quercus Publishing Plc
21 Bloomsbury Square
London
WC1A 2NS

First published in 2009

A catalogue record of this book is available from the British Library

ISBN 978 1 84916 078 0

Printed and bound in China

10 9 8 7 6 5 4 3 2 1

Created for Quercus by Tall Tree Ltd
Designers: Ed Simkins, Sandra Perry
Indexer: Chris Bernstein

Front cover, 3 Denis Scott/Corbis; 5 Dejan750/Dreamstime.com; 10–11 Enjoylife2.../Dreamstime.com; 13 Jfybel/Dreamstime.com; 14–15 Ecoscene/Phillip Colla; 16–17 Ecoscene/Reinhard Dirscherl; 18–19 Ecoscene/Phillip Colla; 20–21 Ecoscene/John Lewis; 22–23 Ecoscene/Reinhard Dirscherl; 24–25 Ecoscene/Jeff Collett; 26–27 Ecoscene/John Lewis; 28–29 Ecoscene/Reinhard Dirscherl; 30–31 Doug Perrine/naturepl.com; 32–33 Sue Daly/naturepl.com; 34–35 Ecoscene/Reinhard Dirscherl; 36–37 Jeff Rotman/naturepl. com; 38–39 Ecoscene/Reinhard Dirscherl; 41 Nataq/Dreamstime. com; 42–43 Ecoscene/Reinhard Dirscherl; 44–45, 46–47 Ecoscene/ Phillip Colla; 48–49, 50–51 Ecoscene/Reinhard Dirscherl; 52–53 Ecoscene/John Lewis; 54–55 Ecoscene/John Liddiard; 56–57 David Fleetham/Oxford Scientific; 58–59, 60–61, 62–63, 64–65 Ecoscene/Reinhard Dirscherl; 66–67 Ecoscene/Jeff Collett; 68–69, 70–71 Ecoscene/Reinhard Dirscherl; 72–73 Ecoscene/ John Lewis; 74–75 Ecoscene/Reinhard Dirscherl; 77 Alexander Hafemann/iStockphoto.com; 78–79 Louie Psihoyos/Corbis; 80–81 Bill Curtsinger/National Geographic/Getty Images; 82–83 Ecoscene/John Liddiard; 84–85 Paul Nicklen/National Geographic/Getty Images; 86–87 Doug Allan/Tartan Dragon/ Oxford Scientific; 88–89 Bill Curtsinger/National Geographic/ Getty Images; 90–91 Juniors Bildarchiv/Oxford Scientific; 92–93 Ecoscene/Reinhard Dirscherl; 94–95 Dorling Kindersley/ Getty Images; 96–97 Alexeys/Dreamstime.com; 99 Dejan750/ Dreamstime.com; 100–101 Ecoscene/Reinhard Dirscherl; 102–103 Ecoscene/Phillip Colla; 104–105 Ecoscene/Reinhard Dirscherl; 106–107 Ecoscene/Phillip Colla; 108–109 Doc White/ naturepl.com; 110–111 Ecoscene/Reinhard Dirscherl; 112–113 Jurgen Freund/naturepl.com; 114–115 Denis Scott/Corbis; 116–117, 118–119 Ecoscene/Phillip Colla; 121 David Shale/naturepl.com; 122–123 Brandon Cole/naturepl.com; 124–125 David Shale/ naturepl.com; 126–127 Ecoscene/Phillip Colla; 128–129 Paul A. Zahl/National Geographic/Getty Images; 130–131 David Shale/ naturepl.com; 132–133 Darlyne A. Murawski/National Geographic/ Getty Images; 134–135 David Shale/naturepl.com; 136–137 Doc White/naturepl.com; 138–139 Dorling Kindersley/Getty Images.